Combat Zoning

Combat
Zoning

MILITARY
LAND-USE
PLANNING
IN NEVADA

David Loomis

University of Nevada Press
Reno • Las Vegas • London

The paper used in this book meets the requirements
of American National Standard for Information
Sciences—Permanence of Paper for Printed Library
Materials, ANSI Z39.48.1984. Binding materials
were chosen for strength and durability.

Library of Congress Cataloging-in-Publication Data
Loomis, David, 1955–
    Combat zoning : military land-use planning in
Nevada / by David Loomis.
            p.   cm.
    Includes bibliographical references and index.
    ISBN 0-87417-187-3 (alk. paper)
    1. Military bases—Nevada—History—20th century.
    2. Public lands—Nevada—History—20th century.
    3. Land use—Nevada—Planning—History
—20th century. I. Title.
    UA26.N4L66   1992
    355.7'09793—dc20

University of Nevada Press, Reno, Nevada 89557 USA
Copyright © 1993 University of Nevada Press
All rights reserved
Book design by Ann Lowe
Printed in the United States of America

9 8 7 6 5 4 3 2 1

CONTENTS

PREFACE

The 1980s were characterized by an unprecedented peacetime military buildup. While the buildup focused on new weapons and increases in force structure, it also involved new demands for land. Nowhere has the military's demand for land been more of an issue than in the deserts of Nevada.

This book began as a paper written for Dr. Earl Kersten's Geography of Nevada course at the University of Nevada, Reno, in the spring of 1987. It is the result of a study of planning for military ranges. This includes such land uses as the Navy's "Bravo" bombing ranges, Air Force missile sites, and National Guard tank training centers. Military land use planning is controversial and has always been a highly visible issue in Nevada and other western states.

Military ranges involve the intensive and often destructive use of more than 16 million acres of rural lands nationwide, including more than 4 million acres in Nevada alone (Map 1). Many of these ranges include intensive air space use as well. In Nevada, approximately 40 percent of the air space is designated for military use (Map 2).

The military expansion initiated by Ronald Reagan in 1982 has resulted in more intense use of existing ranges and more demand for new land and air space. Controversial proposals for land and air space acquisition were developed by various branches of the armed services in California, Idaho, Kansas, Maine, Montana, North Carolina, and Utah. Nevada was singled out for a number of

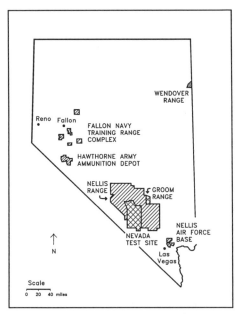

Map 1. Military Land Withdrawals in Nevada

projects because of its sparse backcountry population, vast federal acreage, and favorable climate. In a briefing paper to acquaint Governor Robert Miller with military issues in Nevada, his staff identified the following Department of Defense activity between 1984 and 1989:

9/13/84.   Air Force Gandy Range Supersonic Operations Area established in northeastern Nevada.

12/1/84.   Navy Supersonic Operations Area proposed over central Nevada.

2/1/85.   Navy draft EIS for 181,000-acre Master Land Withdrawal in Churchill County for expansion of bombing ranges.

5/2/85.   Air Force proposal for the Groom Range withdrawal of 89,000 acres in Lincoln County.

5/2/85.   Air Force proposal for Midgetman Missile systems at Nellis and Fallon.

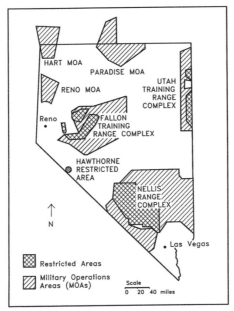

Map 2. *Military Airspace Reservations
in Nevada*

9/1/85.   Air Force Strategic Air Command low-level training route established over central Nevada.

5/2/86.   Hart Military Operating Area approved for the Oregon Air National Guard in northern Washoe County.

7/17/86.   Air Force proposal to locate a laser ranging and
tracking system at Sand Springs Valley in Lincoln County.

10/10/86.   Aerojet land swap of 54,000 acres for testing ranges
for military ordnance and equipment.

7/17/87.   Nevada National Guard request for a 4,523-acre
armory range in Las Vegas Valley.

11/20/87.   Air Force Cruise Missile Route proposal near Tonopah.

7/29/88.   Air Force proposal to increase air space restrictions over the Nevada Test Site.

11/25/88.   Navy housing expansion at Fallon.

11/30/88.   Air Force proposal for a 450,000-acre electronic
warfare range in Utah near Wendover.

1/6/89.   Army National Guard proposal for a 750,000-acre
training range near Hawthorne.

2/1/89.   Navy low-level training route proposed over
Walker Lake and Hawthorne.[1]

Nevadans are a patriotic people. They generally show strong
support for the military and are willing to contribute their fair share
to the nation's defense. Yet they are also are a fiercely independent
people who often object to outside influences.

Because of the high percentage of public lands in their state
(over 80 percent), Nevadans are particularly sensitive to changes
in the status of those lands. Any changes that restrict access to
those lands for economic or recreation uses are questioned. Nevada is the fastest growing state in the nation, and this growth
has led to increased demands on its public lands, particularly for
backcountry recreation. Nevada's backcountry is no longer "the
land that nobody wanted." Management of that land will become
increasingly important in the years to come. Consequently, the
magnitude of Department of Defense proposals for Nevada's land
and air space during the 1980s was cause for considerable concern.

These proposals were not instigated by local citizens. They
have led to a perception that Nevadans have little say in decisions
about military activities in their state. The proposals have caused
much controversy and raised public demand for active involvement in the planning process for military ranges. They have increased participation in interest groups with concerns about the
military's use of land. The proposals have raised many questions
about how the armed services determine their need for land and
air space.

This book will provide background on past military activity in
Nevada, describe current proposals, and analyze one case study
in detail. It traces the military presence in Nevada from John Frémont's 1845 expedition to current Navy proposals to use public
lands to expand its Bravo bombing ranges. The Navy's 181,000-
acre Fallon Master Land Withdrawal is analyzed in detail. That
proposal was initiated more than ten years ago, but is still in the
review process.

Most of the military plans discussed were developed after the passage of the Engle Act and the National Environmental Policy Act (NEPA). The Engle Act transferred authority to approve military land withdrawals from the executive to the legislative branch of government. NEPA has provided citizens with formal and direct access to the military land use decision-making process. Yet it has limitations; formal public input is often required only after proposals have been developed.

The primary theme that emerges from this study is that a lack of citizen participation in the original development of plans is a weakness in the military land use planning process. All citizens have the right to participate in decisions about the use of their public lands. The military should encourage their involvement.

This book is intended for a general audience interested in military history and public land use policy. It is relevant not only to military planning, but also to general public land management because of the impact of military use of these lands and the opportunity costs to other public land users. This book is also intended to be of interest to those in the land use planning field. It illustrates an undeveloped aspect of land use planning that has been on the front page of newspapers throughout the West for more than a decade.

Traditionally, military land use planning has been concerned with the urban planning aspects of military bases, particularly housing. It has become so well developed that it now has its own division of the American Planning Association—Federal Installation Planning.

The literature of planning includes extensive analysis of land uses that involve public nuisances, especially adult theaters and halfway houses. It does not include any analysis of military ranges. This oversight is because military ranges generally do not affect urban areas. But the armed services' demands for land during the last decade point to a need for the planning profession to turn its attention to military ranges.

My sincere thanks to those who have helped with this project. Dr. Earl Kersten of the University of Nevada, Reno, helped me formulate the ideas that led to this book. Senator Richard Bryan took time out of his busy schedule to discuss military planning with me. The senator is clearly a public servant who knows the meaning of public service. Drs. Chris Exline, Al Wilcox, and Gary Housladen of the University of Nevada, Reno, provided invaluable reviews of the text and the encouragement needed to pursue this project.

Robin Datel and her colleagues with the Association of Pacific Coast Geographers helped focus the policy analysis on geographic perspective.

Stan Sloss, counsel for the House National Parks and Public Lands Subcommittee provided timely documents. Don Bradley, formerly of the Naval Facilities Engineering Command, Jim Regan, Churchill County commissioner, Grace Bukowski of Citizen Alert, and Cindy Wood of the University of Nevada Press were also most helpful.

The research for this study was made much easier by the library staffs of the University of Nevada, Reno; the University of Nevada, Las Vegas; Stanford University; the University of California, Berkeley; and especially Allison Cowgill, Annie Kelly, and the rest of the staff at the Nevada State Library in Carson City.

My wife, Melissa, contributed the editing skills she has perfected at *Nevada Magazine* and sustained the author with her patience and understanding.

The viewpoints and conclusions expressed in this book are solely the author's—as are any omissions and errors.

ONE

Battle Born

The military has always played a significant role in Nevada. Even before its admission to the Union, the United States Army had taken an active interest in the area. The rising tide of "manifest destiny" in the 1830s and 1840s prompted the Army to mount several major expeditions to Nevada. Although these expeditions were primarily intended for exploration and mapping, military motives were also significant. Captain John C. Frémont's 1845 expedition was ostensibly to discover an efficient route from the Mississippi River to the Pacific. He explored the Humboldt Basin, Ruby Mountains, Big Smokey Valley, Walker Lake, and Donner Pass. His explorations generated public interest in the Great Basin and added to the scientific knowledge of the area's geography and natural history. However, Frémont noted that "in arranging this expedition, the eventualities of war were taken into consideration."[1]

The expedition played a key role in the nation's manifest destiny to span the continent. The war with Mexico followed soon after, and in 1848 the federal government purchased the territory that was to become Nevada from Mexico.

The military's role in the exploration of the new territory was significant. Major expeditions by the Army Corps of Topographical Engineers led by Lieutenants Edward G. Beckwith and Joseph Christmas Ives, Lieutenant Colonel Edward J. Steptoe, and Captain James H. Simpson added considerably to public knowledge of both northern and southern Nevada. The Army's exploration efforts culminated with Captain Simpson's expedition of 1859 to find a shorter wagon route from Salt Lake City to San Francisco. He explored much of central Nevada and laid out the general route of what was to become U.S. Highway 50.[2]

The attainment of statehood was also the result of military activity—the Civil War. Nevada's silver and gold production helped finance in large part the Union's war effort. Also, in 1864, President Lincoln expected a major reelection battle with the peace faction that wished to negotiate with the Confederacy. In order to gain its electoral votes, Lincoln obtained statehood for Nevada, thus the state motto, "Battle Born."

A clause in the act of Congress that authorized statehood contained a provision that later played a key part in military planning for the state. This required the citizens of the new state: "To forever disclaim all right and title to the unappropriated public lands lying within said territory, and that the same shall be and remain at the sole and entire disposition of the United States."[3]

This clause was also required of all neighboring states upon their entry to the Union. The federal government originally granted Nevada 3.9 million acres of land to provide funding for schools. The land consisted of the 16th and 36th sections of each township. Realizing that most of this land was almost worthless at the time, Nevada relinquished its claim and in return received about 2 million acres selected at its choice. This swap netted the state lands much more valuable than the lands it relinquished—a far better grant than the other western states received. Unlike most of these states, however, Nevada sold most of its land to developers. Today, the state owns only 130,000 acres of the original grant. The swap probably led to improved land management since most western state governments do a poor job of managing their land grants. But it did make future military withdrawals much easier, since remaining federal lands were in larger solid blocks.

Meanwhile, many military posts had been established in Nevada to protect settlers against Native American uprisings. The most famous of these occurred in 1860. The Pyramid Lake War began with the capture of two Native American women by whites from Williams Station. The natives responded by killing three whites and burning the station. After a disastrous battle involving volunteers from Virginia City, the regular Army settled the war and established Fort Churchill on the Carson River. The Army had earlier built two smaller forts, Haven and Storey, along the Truckee River. It selected the Churchill site for the major installation due to its location midway between the Pyramid Lake and Walker River Paiutes.[4] Fort Churchill served as military headquarters for Nevada until its abandonment in 1869.

In all, there were twelve military installations in the northern part of the state to police the Native Americans. A larger number of temporary camps were set up at various other points during local emergencies. Fort Ruby was located on the southern end of the Ruby Mountains in White Pine County. It was established in 1862 to protect the Overland Mail Route. The fort was involved in major hostilities with the Goshute, Shoshone, and Paiute tribes from 1863 to 1865. But its long-term impact was more peaceful: Fort Ruby created a local demand for livestock, hay, grain, and produce and thus helped develop agriculture in northeastern Nevada. Many discharged soldiers settled in the area and turned to farming and livestock ranching.

Fort McDermitt was the longest lived of any Nevada post. Established in 1865 in northern Humboldt County, it protected stage routes and wagon roads. Later, its main purpose was to protect the nearby Indian agency. It was finally closed and turned over to the Department of the Interior for use by the Indian agency. Fort Halleck was active on the northern slope of the Ruby Mountains from 1867 to 1886. It was established to protect the settlers in the Humboldt River area and the route of the Central Pacific Railroad. Two other installations of note—Camps McGarry and Winfield Scott—were located in Humboldt County. Their mission was to protect the emigrant routes to California and Oregon. These and other camps contributed to the military security of their areas but were short lived.

The Army Corps of Engineers remained active during this period. Survey parties led by Fort Halleck's Lieutenant George M. Wheeler made important contributions to the knowledge of Nevada geography. With the pacification of the Native Americans and the coming of the railroad, locally stationed soldiers were no longer needed and the Army eventually closed all of its posts. It abandoned Fort Churchill in 1869 and Fort McDermitt in 1888.[5] The closure of Fort McDermitt essentially ended the state's relationship with the military for nearly forty years.

TWO

Bombs, Politics, and Hunting at Hawthorne

I n July 1926, a fire and explosion at the Navy's Lake Denmark, New Jersey, ammunition depot killed fifty people and injured hundreds more in nearby towns. This prompted Navy planners to search for a new site. A court inquiry into the fire required them to find enough land to allow for the safe grouping of ammunition magazines to prevent a repeat of the Lake Denmark disaster.[1]

Navy planners also wanted an inland site for storage away from invasion areas on the coast and away from any population centers. They narrowed their selections to two choices: Hawthorne, in Mineral County, Nevada, and Herlong, in Lassen County, California. Both sites had the advantage of already being owned by the federal government. At the time, Nevada Senator Tasker Oddie was a ranking member of the Naval Affairs Committee. He prepared a report that concluded that the proposed construction of the Golden Gate Bridge would interfere with the movement of the fleet in time of emergency. He claimed he opposed the bridge because it would create public fears about the Navy's ability to respond to military threats. When California Senator Hiram Johnson demanded an explanation, Oddie replied:

There'd probably be less fear if they knew that they had a tremendous backup for the fleet in the form of a new, well-established ammunition depot well behind the Sierra Nevada mountains, which could give strength to the Navy and all branches of service in time of war.[2]

Senator Johnson saw the wisdom in this argument and soon
Nevada had its ammunition depot and California its bridge. This
established a process in common with local government planning—professional staff developed alternatives, but final site selection was purely political. This process was to become codified
into federal law in 1958 with the Engle Act.

The land selected for the ammunition depot consisted primarily of public domain lands owned by the federal government under
the jurisdiction of the secretary of the interior. The executive
branch could have reserved or withdrawn lands for the Navy under
two authorities. The first was the implied authority of the executive. In the case of the *United States* vs *Midwest Oil Company*
(326 U.S. 459, 474), the Supreme Court's opinion was that although
Congress has control over the public domain, the power of withdrawal by the president as an agent had been used over a long
period without objection by the Congress. The second alternative
was the authority granted by the Pickett Act of 1912. It authorized
the president to *temporarily* withdraw any of the public lands for
waterpower sites, irrigation, or "other public purposes."[3]

President Coolidge decided to exercise his authority under
the Pickett Act for the Hawthorne plant. By executive order, he
reserved lands "for the exclusive use and benefit of the United
States Navy for the development of and use as an ammunition
depot, until this order is revoked by the President or Congress."[4]
Under the Pickett Act, this was an acceptable practice for a temporary withdrawal. Subsequent executive orders adjusted the boundaries to their present configuration, encompassing an area of
147,000 acres.

Nevada welcomed the depot, particularly because the state's
economy was suffering from a decline in mining activity. Federal
public domain lands were considered worthless. They were available for any productive use.

The commissioning of the Naval Ammunition Depot at Hawthorne opened a new era in state/federal relations in Nevada.
The depot, ultimately covering some three hundred square
miles of desert wasteland, showed the way to an effective use
of part of Nevada's sparsely settled, wide-open spaces and

established a precedent for future federal programs during and
after World War II.[5]

The depot has two distinct areas. About two-thirds consists of
saltbush flats east of Hawthorne. This area is used for ammunition
storage and other military operations. The remaining third is a
scenic mountainous area west of Hawthorne dominated by Mt.
Grant (11,239 ft.). The Navy originally intended to use the Mt.
Grant area for watershed and security purposes. In the thirties, it
built three small reservoirs and constructed a system of catch bas-
ins to maximize the area's water yield. The system is still in use
today. The Navy considered security on Mt. Grant important before and during World War II, but with today's satellite technology, spying from the ground is no longer an issue.

Mt. Grant has also been used for other purposes. In 1931 a
summer camp was established for base personnel. They enjoyed
private access to the area for hunting and fishing,[6] a practice that
later became controversial:

During World War II the withdrawn area became a private hunting preserve for Naval officers and VIP civilians. In
the mid-1950s there were complaints of poaching and an officer was arrested for shooting grouse out of season. In an effort
to improve public relations, hunting was permitted in the area
on holidays and weekends on a first come first served basis by
permit only. Usually base personnel got most of the permits.[7]

The Hawthorne operation is the largest ammunition depot in
the world.[8] It started in the early thirties with 160 workers refur-
bishing Navy mines from World War I. This activity dwindled until
1940 when employment dropped to 100. World War II revived the
depot, employment reached its peak in 1945 with more than 5,000
workers. It was the largest staging area for bombs, rockets, and
ammunition for the entire war effort.

Activity at the depot peaked again during the Korean and
Vietnam wars. In 1977, the Department of Defense centralized
all ammunition functions under the Army. This economy effort
involved transferring the Hawthorne depot from the Navy to the
Army. In 1980, the Hawthorne Army Ammunition Plant was con-

verted to a contractor operation by the Day, Zimmerman, and Basil Corporation.

The ammunition plant is by far the largest employer in Mineral County. Of the county's 6,300 residents, about 2,000 are directly or indirectly associated with the plant.[9] The remaining population is primarily supported by farming on the Walker River Indian Reservation, tourism at Hawthorne and Walker Lake, and scattered mining activity in the county's rugged scrub pinyon- and juniper-mantled mountain ranges.

Current activity at the ammunition plant involves the storage, production, maintenance, testing, and demilitarizing of munitions. A primary job is destruction of obsolete 20 mm ammunition. The plant has nearly 2,000 earth-covered bunkers capable of storing 500,000 tons of ammunition. Other storage is provided for NASA rockets and Minutemen and Polaris missiles. The plant provided munitions from small arms fire to 2,000-pound bombs for the Persian Gulf War to the Army, Air Force, Navy, and Marines.

The Navy is still active at Hawthorne, operating a naval mine assembly facility on site. Since 1987, the Strike Warfare Center at the Fallon Naval Air Station has used the plant as a no-drop bombing site. It is used as a simulated urban area for nighttime bombing practice.[10]

THREE

World War II Bases

orld War II not only affected the Hawthorne depot, but began an extensive military presence in Nevada that has continued to this day. Fear of a Japanese invasion of the West Coast led to the establishment of military bases throughout the state. This included Army Air Corps bases at Tonopah, Stead, Minden, Lovelock, Wendover, and Winnemucca. The Minden and Lovelock bases were closed following the Battle of Midway, when the Japanese lost their naval superiority.[1] Nearly 25,000 airmen trained at Wendover, including the crews that dropped the atomic bombs on Hiroshima and Nagasaki. The base closed after the war and was sold to the city of Wendover for one dollar.[2] The Tonopah base has been used intermittently with the Nellis Range and served as the home of the nation's only Stealth fighter wing. After the war, Stead Air Base became the site of the Air Force's survival training school, a helicopter training program, and a major radar installation. Its closure in 1965 had a major impact on the local economy as did the closure of the Winnemucca Air Force Station in 1968.[3] Since then, Stead has been used by National Guard and Reserve units as well as the Reno Air Show. These facilities have been relatively insignificant statewide, compared to the establishment of the Nellis and Fallon ranges.

The Nellis Range

On October 29, 1940, President Roosevelt established the Las Vegas Bombing and Gunnery Range. Now called the Nellis Range, it is the largest military range in the Western world. Executive Order 8578 established the range under an obscure legal authority granted by the Army Appropriations Act of July 9, 1918. It provided for allowing the president to reserve unappropriated public domain lands for aviation fields for testing and experimental work.

The original executive order reserved 3.5 million acres for the range. It partially overlapped the Desert National Wildlife Range established by Roosevelt in 1936. Since then it has been subject to many other executive orders, public land orders, and public laws that have modified its boundaries to the current 3 million acres.

The Nellis Range is typical of the Basin and Range physiographic province, characterized by north-south–trending mountains flanked by alluvial valleys. Major topographic features are the Pahute Mesa uplands, the Pintwater, Belted, and Kawich mountain ranges and the dry lake valleys of Gold, Cactus, and Yucca flats and Kawich Valley. Much of the Nellis Range is composed of the salt desert shrub community of shadscale and greasewood. The southern portions contain extensive stands of Joshua tree, yucca, prickly pear cactus, and creosote bush. Wild horses and bighorn sheep are found throughout the range. Primary land uses before the withdrawal were ranching and mining.

The Army Air Corps selected the area for its gunnery school based on the following planning criteria:

> The weather is excellent for flying and there was plenty of government land available for a dollar an acre. The land was cheap because it really wasn't much good for anything but gunnery practice—you could bomb it into oblivion and never notice the difference.[4]

Actually, the land didn't even cost a dollar an acre. It was free, being already owned by the federal government. The price did not reflect its true value. Even in the forties, there were objections from others who valued the land. Though most Nevadans gave it little thought, ranchers and miners who used the area complained. But that was during wartime, and few took the withdrawal very seriously.[5]

Nellis Air Force Base was deactivated in 1947 following the

end of the war. The newly created U.S. Air Force reactivated it in 1949 due to increasing tensions in Korea.

In 1951, 435,000 acres were deleted from the Nellis Range for the establishment of the Nevada Nuclear Test Site, now managed by the Department of Energy. The Navy made requests in 1953 and 1955 for joint use of the remaining Nellis Range for training. Air Force officials rejected these requests because the range was being fully used by its own aircraft. That conclusion was questionable, however, considering the Air Force's decision in 1956 that most of the range was no longer needed. Meanwhile, the Navy pursued other options and claimed that the Air Force reversal had come too late to change its plans. Congress saw this as a failure by the Department of Defense to coordinate effectively the land use planning activities of the armed services.

The repeated Air Force turndowns of Navy requests for joint Air Force-Navy use of Nellis-Tonopah, and defense of that action in January 1956, only to announce release of more than 2 million acres there 2 months later, was labeled by this committee in its 1956 report "inexplicable." The Navy's plea that the . . . release came "too late," was similarly labeled "incomprehensible." Finally the action of Defense in blessing all of the positions of both departments was labeled "an inexcusable deficiency in control procedures."[6]

Congressional pressure eventually led to joint Air Force-Navy use, although management of the Nellis Range remained solely with the Air Force. The area released by the Air Force was later reduced from 2 million acres to less than 200,000.

The Air Force generally had good relations with local residents, but two events suggested the potential for public controversy and demonstrated some of the problems currently facing the military in Nevada. Although the Air Force had identified most of the range as unnecessary, in 1956 Nellis officials wanted to expand into an area populated by bighorn sheep. Local hunters strongly opposed the proposal, and the controversy eventually led to the proposal's demise. Soon after, ranchers fought Air Force proposals to eliminate livestock grazing on the Nellis Range. This resulted in a settlement in which taxpayers paid for property they already owned.[7]

The Air Force had permitted ranching and mining on portions of the Nellis Range since its inception. It eventually decided that such uses were incompatible with military activities. In 1956, the Air Force paid $708,000 to revoke mining claims and livestock grazing privileges. Although the purchase of mining claims by the Air Force was legal, the purchase of grazing privileges was questionable. No legal property rights had been assigned to ranchers for livestock grazing on federal lands.[8] This meant that in paying to revoke grazing privileges, the Air Force was buying a property right the federal government already owned.

The Nellis Range was used extensively for training pilots for the Korean and Vietnam wars. Pilot losses during those conflicts pointed to the necessity of increasing the realism of air combat training. This resulted in the development of the "Red Flag" training program. In this training program, Air Force aggressor squadrons electronically simulate attacks by enemy aircraft using Soviet and other potential enemy air-to-air combat techniques. Trainees also face electronic warfare sites that jam radar equipment and simulate surface-to-air missiles. Army, Navy, National Guard, and Air Force Reserve pilots also participate in the training, as do pilots from almost every U.S. ally.[9]

Nellis is well suited for training for combat in the Middle East due to similarities in climate. Visual recognition remains an important factor in pilot training. The aggressor squadrons have used Soviet desert camouflage to improve the trainee's visual skills.

The object of Red Flag exercises are "Alpha Strikes" and low-level bombing runs on enemy ground installations. These are full-fledged missions involving both fighter and bomber aircraft. The Nellis Range has more than fifty different types of enemy targets including simulated tanks, missile sites, antiaircraft guns, runways, helipads, industrial sites, bridges, radar sites, rail yards, trains, tunnels, and pipelines. The missile sites include simulated surface-to-air missiles called "Smoky SAMs." These were developed after Red Flag pilots complained that the simulated Soviet missile sites had an unfair advantage. Pilots couldn't evade those sites very well without visual cueing from a smoking missile. The

Smoky SAM, which trails 1,000 feet of white smoke, evened the odds.

The Red Flag program trains for attitude as well as skill. The ideal attitude for a fighter pilot is: "Hell yes, I'm the best there is and everybody else better keep out of my sky." Those that successfully attain the proper attitude are "the real killers-of-the-sky."[10] With this kind of attitude, the concerns of local residents are probably not a high priority, as is evidenced by complaints from nearby communities about "buzzing" and sonic booms over their homes.

The Nellis Air Force Base is a major contributor to the economy of southern Nevada. It accounts for nearly a billion dollars of gross regional product. Direct and indirect employees and their dependents amount to almost 62,000 residents or about 8 percent of the population of Clark County.[11] It provides much-needed diversification for an economy dominated by the casinos in Las Vegas.

The Nellis Range has been managed under a five-party cooperative agreement originally signed in 1977. It involves the Air Force, Bureau of Land Management (BLM), Nevada Department of Wildlife, U.S. Fish and Wildlife Service, and Department of Energy. The agreement provides for the protection and management of wildlife, watersheds, and wild horses on the Nellis Range and the Nevada Nuclear Test Site. It focuses on the area jointly managed by the U.S. Fish and Wildlife Service as part of the Desert National Wildlife Range and the area managed by the Bureau of Land Management as the Nevada Wild Horse Range.

The Military Lands Withdrawal Act of 1986 renewed the withdrawal of the Nellis Range and required the BLM to prepare a resource management plan for the range. Up-front public participation in its development was an important feature of the plan. In July 1988, the bureau sent a scoping report to 250 individuals, interest groups, and local, state, and federal agencies. Scoping is a process outlined by the National Environmental Policy Act that provides the public an opportunity to define the scope of environmental impact statements (EIS). The report summarized the BLM's management and resource issues and asked the public to review them and identify their own issues. Public meetings were held in Alamo, Tonopah, and Las Vegas to explain the planning process and discuss citizen concerns. Only fifteen people attended the three meetings. Twenty-five others mailed their comments. Public concerns focused on wild horses, desert tortoise habitat, endangered species, mining, cultural resources, wilderness, and public access.[12]

From these concerns, BLM staff focused the plan on four issues: vegetation, wildlife habitat, wild horse and burro management, and cultural resources. The other public concerns were dropped because no area of the range qualified for wilderness designation, and the Air Force had already foreclosed the possibility of any mining or public access to the range.

Planning criteria were then prepared to guide the development of the plan. These included a recognition that the primary purposes of the lands were for pilot training and weapons testing for the Air Force.

The BLM developed a preferred alternative that included proposals to protect riparian vegetation, develop water sources for wildlife, round up wild horses and burros to keep numbers within the carrying capacity of the range, and protect cultural resources.

The bureau released a draft plan and an environmental impact statement for public comment in May 1989. It held public hearings in Alamo, Caliente, Tonopah, and Las Vegas in July. Again attendance was light and little controversy was evident. Most of the comments on the plan concerned wild horse management. They were addressed in a final EIS issued in January 1990.

The Fallon Ranges

World War II also led to the creation of Nevada's second largest military base, the Fallon Naval Air Station. It was first established as an Army Air Corps field in 1942. In 1944, the Army turned it over to the Navy as an auxiliary air station for the Alameda Naval Air Center. Its mission was to provide training to Navy air groups. After World War II it was placed in caretaker status and transferred to the Bureau of Indian Affairs, only to be permanently reactivated during the Korean War.

The on-again, off-again Black Rock and Sahwave ranges (Map 3) led to the first major statewide controversy about military activ-

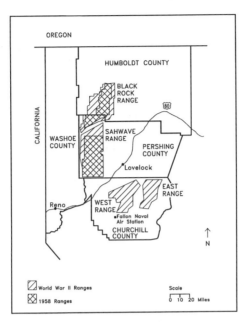

*Map 3. Early Navy Bombing and
Gunnery Ranges*

ity. These ranges were key in establishing a national policy about
military planning and provided valuable lessons about public involvement in the planning process.

The Black Rock Desert is 100 miles northeast of Reno. It is a
vast, dry alkali lake bed totaling 500,000 acres. It is part of the
remains of ancient Lake Lahontan, which once covered much of
western Nevada. The Black Rock Desert is one of the largest undeveloped desert valley floors in the nation. It contains the Quinn
River drainage, an ephemeral stream often dry by late summer. It
was once used by pioneer emigrants to California. Part of the desert
is on the National Register of Historic Places. Recent discoveries
include the remains of a woolly mammoth, a saber-toothed tiger,
and other Pleistocene animals that may have been trapped by the
marsh shoreline of Lake Lahontan. Today, antelope, deer, cougar,
sage grouse, and chukar partridge are found along the fringes of
the Black Rock Desert. Beaver and muskrat have been spotted
along the Quinn River. The desert is also used by rock hounds,
history buffs, and others who enjoy recreating there.

The Sahwave Mountains area includes a series of north-south–trending mountain ranges. Upper elevations are snow covered in winter and have scattered pinyon and juniper forests. Several drainages have stands of aspen and high mountain meadows. Lower elevations have extensive stands of sagebrush, saltbush, and greasewood. Mule deer, wild horses, and burros are found throughout the area. The Sahwave Range has been used for years for mining, livestock grazing, and hunting.

In 1942, a withdrawal order granted the Army Air Corps use of 623,000 acres in Pershing and Humboldt counties for use as the Black Rock bombing range. The Department of the Interior canceled the range in 1943. Under authority granted by the secretary of the Navy, more than 1 million acres were closed to the public in October 1944. This area, the Lovelock Aerial Gunnery Range, consisted of two sections. The North Range totaled 700,000 acres in the Black Rock Desert area. The South Range amounted to more than 800,000 acres in the Sahwave Mountains area.

The Navy closed the range to mining, grazing, and all other land uses, warning that live ammunition would be used in air-to-air and air-to-ground gunnery. It gave residents one month to remove themselves and their belongings from the range or face prosecution.[13] In February 1945, the Department of the Interior issued a legal permit for use of the Lovelock Range for air-to-air gunnery training only. The permit provided that the range would be in effect only until six months after the national emergency declared for World War II was canceled.

The Navy relinquished the South Range in 1946. In March 1949, it requested permission from Interior's newly created Bureau of Land Management to replace the North Lovelock Range. The Navy wanted a new Black Rock Range near the town of Sulphur. The new range was to be used for both air-to-air gunnery and air-to-ground bombing training sites for carrier groups from the Fallon Naval Air Station.

By May 1949, the BLM and a group of local livestock owners had worked out the boundaries of a range totaling 272,000 acres in cooperation with the Navy. In September, the Department of the Interior issued a letter granting the Navy use of the range for five years.

In 1953, the Navy decided to reactivate the South Lovelock Range. It applied for a permanent withdrawal of 541,000 acres for air-to-air gunnery, this time calling it the Sahwave Range. Two years later, it applied for permanent withdrawal of the Black Rock Range and as well for an addition to the Black Rock Range of 1.4 million acres and an addition to the Sahwave Range proposal of 655,000 acres. This brought their total requests to 2.8 million acres.

These applications were filed under Executive Order 10355 (May 26, 1952), which provided that the lands would be protected from appropriation by miners under the 1872 Mining Law. The lands were to remain in public ownership until the secretary of the interior made final decisions on the applications.

In its justification, the Navy indicated that the withdrawals were necessary for the national defense. It claimed the need for one thirty-by-fifty-mile area (1.2 million acres) in the Black Rock Desert and one twenty-three-by-fifty-mile area (1 million acres) in the Sahwave Mountain area. The Navy said that it aligned the two rectangular areas within a larger area to maximize efficiency. It needed 600,000 more acres for "triangles and zig-zags" in the range boundaries, to follow section lines, and to avoid mining claims.[14]

The ranges were for air-to-air gunnery only; no bombs, rockets, or explosives were to be used. Pilots were to train by firing their machine guns at targets towed by other aircraft at high altitudes. The targets would be towed at the end of 2,000 foot cables at 200 miles an hour. According to the Navy, each area would have five separate targets towed at once with planes darting in from the outside to fire at them. The ranges were needed only to contain falling bullets. The Navy maintained that although a person standing in the range probably would be unaware that firing was being conducted, the hazard from falling bullets justified closing the ranges to public use.

The withdrawal requests led to 243 public protests, primarily from grazing, mining, and wildlife interests. The general public was concerned about the overall amount of public land used by the military. The following protests were typical.

We have read and studied the arguments on both sides, in the Navy's procedure to take over a large section of Northern

Nevada. We still feel this land is invaluable as a recreational area, not only for Nevadans but for everyone. The Military has enough of the West already. Our increasing population makes it more than ever important that we preserve our wildlife and wilderness areas for future generations.[15]

What appears from a jet cockpit to be wasteland can look very valuable at ground level from a miner's or a stockman's point of view. Withdrawal of any land that can be used for mining or agriculture is a definite harm to the State of Nevada at large.[16]

Public reaction to these and other applications for military land withdrawals caught the attention of Congress. In October 1955, Representative Clair Engle of California, chairman of the House Interior and Insular Affairs Committee, asked the secretary of the interior to withhold approval of any further withdrawals of public land for military purposes until his committee investigated the matter. The secretary agreed.

While Congress studied the withdrawal issue, controversy in Nevada continued. The Navy Black Rock and Sahwave proposals became the number one issue in the state during 1956 and 1957. In response to public questioning about the need for the ranges, the Navy made a strong case based on the development of high-performance jet aircraft. Their high speed and wide turning radius required larger training areas and more intensive pilot training. The Navy made a credible case for the need to triple its training missions on the West Coast to qualify pilots in aerial gunnery.

The Navy was less successful in explaining why it could not use alternatives to the Black Rock and Sahwave ranges. In response to questions about whether it could increase training over the ocean, the Navy replied that it had difficulty controlling civilian shipping under training areas. Yet it gave no evidence of any safety problems that its training over the ocean created for civilian ships. The Navy also pointed out that training missions off the Pacific Coast could only be launched on half as many days due to poor weather conditions. Nevadans questioned the effectiveness of training pilots only in their state's bright sunshine and remained unconvinced that the Navy should avoid training its pilots in all kinds of weather conditions.[17]

Nevada's preferred alternative was joint Air Force-Navy use of the Nellis Range. The Air Force's 1955 assertion that it was using the range to full capacity proved incorrect. A reevaluation the following year led to a conclusion that 2 million acres were available for other agencies.

The Navy responded that this reevaluation came too late to meet its needs. In order to use the Nellis Range it needed to rehabilitate the old Tonopah Air Base. This would take at least two years and the need for increased training was immediate. The Tonopah base, on the northern boundary of the Nellis Range, would be used as a refueling station. The Navy threatened to close the Fallon Naval Air Station. If a refueling station was available, the Navy could fly its planes from permanent bases at Alameda and Moffett fields in California.

The Navy also stated that the area the Air Force identified was the wrong shape. It included a ninety-degree bend that prevented it from being used as a gunnery range. Also, the Navy expected that the Atomic Energy Commission probably would ask for the additional acreage. It believed this was preferable because the Air Force had contaminated the range with unexploded ordnance. According to the Navy, it would be more efficient to use the range for some contaminating military operation rather than for the Navy's noncontaminating aerial gunnery training.[18] Nevadans didn't buy these arguments. Neither did Congress. It promised to enact a law requiring an analysis of existing ranges before new ones could be withdrawn.[19]

In attempting to refute the Navy's statements, Nevadans argued that given the speed of its newer aircraft, the Nellis Range was only five minutes further from the Fallon Naval Air Station than the Black Rock and Sahwave ranges. In one of its protests against the Navy proposal, the Pershing County Chamber of Commerce brought out a continuing problem for public participation in military planning. The chamber said that despite its attempt to analyze the Navy plan, it was extremely difficult for it or other citizens to oppose the Navy. The Navy's huge resources included a large staff in Washington with access to Congress and other federal agencies. Its bureaucratic resources, power, and standing with the president were far greater than the other organizations involved in the process. Until 1961, when more centralized systems and procedures were implemented, the Navy operated independently, controlling its own budgets and resource allocations.

The Pershing County Chamber of Commerce also argued that the Navy had a major advantage in working on the plan in secret.[20] The Navy had two years to complete its plan and develop strong rationale without informing the public or providing any opportunity for citizen participation in the planning process. This lack of up-front public input was a major factor in the intensity of opposition to the Navy's plans.

In the face of public resistance, the Navy began to rethink its plans. By March 1957, the Navy announced major changes in its proposal. It eliminated 1.5 million acres from its original 2.8-million-acre request by dropping the Black Rock Range extension entirely and adjusting the boundaries of the Sahwave Range. It was able to make these adjustments by coming to an agreement with the Air Force on the use of the Nellis Range.

Other concessions were made to accommodate public land users. The Navy agreed to purchase mining claims outright or lease existing mines and relinquish them to their current owners when the ranges were no longer needed. It agreed to cease operations for five months of the year and on weekends and holidays to allow access for livestock ranchers and hunters. It further agreed to drop its proposal for permanent withdrawals, instead limiting the withdrawals to five-year periods. Nevada Representative Cliff Young supported the compromise and applauded the Navy's responsible reaction to public concerns. He believed that the Navy had made an unprecedented effort in military-civilian struggles to coexist.[21]

The Navy concessions were included in revised withdrawal petitions to the Bureau of Land Management. Before taking any action on the withdrawal, the bureau held a public hearing in Pershing County. The hearing was held in Lovelock in June 1957. Despite the Navy concessions, public opposition continued. There was unanimous, determined resistance to any withdrawal whatsoever.[22] Much of the disagreement came from miners, who focused on past problems with the Navy.

What I am objecting to most strenuously is that for the past three years we have been tied up, we haven't been able to operate, and the first contact we had with the Navy was very objectionable. They came out and said pack up and move, you have had it, you are done. Well, since being there since 1947, we weren't quite inclined to move quite that rapidly. We intend to stay there until we get some higher authority to move.[23]

Many miners made grandiose claims about the value of their mining properties. One insisted that he had the "largest and finest" perlite mining property in the country, if not the world, though the records didn't show it.

Others fought the withdrawal because they did not believe Navy promises to open the ranges to the public each year.

Arrangements whereby land acquired by the navy is left open at certain times cannot be wholly satisfactory. Such land is subject to immediate closure to all comers if the navy deems such actions feasible. Peacetime promises that gunnery range land will not be peppered with explosives are quickly withdrawn if an emergency threatens, and the land can be ruined for decades or even generations thereby.[24]

Forrest Cooper, an attorney representing the livestock industry, made a lengthy and impassioned argument against the withdrawal, interrupted often by applause from the crowd. Navy Captain H.C. Bridger replied that while he respected Mr. Cooper's speaking skills, "he isn't qualified to say whether we need [the withdrawal] or not." The captain was implying that the Navy's experts were better qualified than citizens to judge what is in the public interest. The view that complex decisions should be made only by technical experts was common in the fifties. It reached its peak in transportation planning, where experts needlessly disrupted entire cities with new highways based solely on efficiency criteria. Many experts believed that decisions should be made based on technical considerations only, free of citizen values. This type of technocratic approach is generally rejected today as antidemocratic.[25]

In its analysis of the hearings testimony, the BLM focused on the mining issue. It decided that the miners were grossly exaggerating the value of their claims to get higher than market value for them from the Navy. If mining properties were indeed as valuable as claimed, far more active mining would have been occurring in the area. Since there was little mining activity and the Navy had made a strong case for the ranges, the Bureau recommended that the withdrawal be approved subject to some minor wording changes based on the hearings testimony.

The exaggerated claims of the miners were harmful to the entire opposition. At the same hearing that the opposition was questioning the credibility of the Navy, its own credibility became a determining factor.

The case for the withdrawal was strong enough to convince Congress to exempt the Black Rock and Sahwave ranges from an act it was debating that would require congressional approval of all military withdrawals. The approval decision for the Black Rock and Sahwave ranges was left to the secretary of the interior. This was a major concern of Pershing County. It was convinced that the Navy was not providing the Department of the Interior with all the information necessary to make a decision on the withdrawal. The county believed that withdrawal procedures prevented the Department from obtaining that information.[26] While the Department of the Interior did not have all the Navy's information, the extensive controversy, congressional investigations, media coverage, and public hearings probably gave it more information on the Black Rock and Sahwave proposals than it had ever had for any previous withdrawals.

Further negotiations trimmed the Navy's request by another 500,000 acres. In May 1958, the secretary of the interior approved the withdrawal. Public Land Order 1632 withdrew the original Black Rock Range of 272,000 acres and the Sahwave Range of 519,000 acres for Navy aerial gunnery training. The withdrawal order was subject to a number of important limitations. It provided that the BLM could issue permits and leases for recreation, wildlife, livestock grazing, and other surface uses. It closed military uses during January, February, March, May, and October of each year, during which time the lands could be used for recreation and access by livestock ranchers to tend their herds. It prohibited any military uses that could result in long-term contamination of the land and limited the withdrawal to five years subject to an additional five-year extension.

The Navy operated the ranges as provided in the public land order until 1963 when the first five-year period ended. The Navy relinquished the Black Rock Range because it was too small for its new faster aircraft. The Black Rock Desert has recently been recommended for designation as a wilderness area. The BLM wilderness analysis noted that although most of the proposed wilderness area had been used as a military range, the only remaining evidence is a few ammunition shells.

The Navy identified the continued need for the Sahwave Range but wanted to change the restrictions so that it could be used year-round. It asked the Bureau of Land Management if it could investigate the possibility of providing other areas for the ranchers that used the range. Navy officials requested that the investigation not be publicized because they were afraid of congressional repercussions. The BLM disagreed and issued press releases on the proposed total closure of the Sahwave Range with a request for public comment.[27]

Public reaction was predictable: Nevadans opposed any further closures of the Sahwave Range. They remembered Navy promises to keep the range open for grazing, hunting, and other recreation and insisted those promises be kept. Pershing County officials were afraid that access restrictions would prevent a proposed solid fuel plant from being constructed by the Aerojet Corporation. They expected Aerojet to provide a major boost to the Pershing County economy. The state strongly opposed year-long closure of the range, citing previous Navy agreements to provide public access for five months of the year for grazing and recreation use.

The Navy reacted responsibly to public concerns. It dropped the request for year-round closure of the Sahwave Range and honored its commitment to provide public access for five months each year. Since this resolved any controversies, the Department of the Interior extended the withdrawal of the Sahwave Range for five years.

By 1965, development of guided missile technology had rendered conventional air-to-air gunnery obsolete. One result was the

Navy's decision to relinquish the Sahwave Range. Although some public pressure was necessary, the Navy kept promises it had made during the withdrawal process in the fifties. Nearly 1 million acres were returned to the Bureau of Land Management and opened to the public. The lands are now managed for public recreation, watershed protection, wildlife habitat, grazing, and mining. All Nevadans benefited.

The Black Rock and Sahwave ranges were not the only areas of controversy for the Fallon Naval Air Station. When the Navy assumed control of the airfield in 1944, it applied to the secretary of the interior for 1.6 million acres of land northeast of Fallon for a bombing and gunnery range. Interior's Grazing Service, forerunner of the Bureau of Land Management, objected to the size of the range because of impacts on the local ranching community. In response, the Navy trimmed its proposal to 700,000 acres in two ranges, one located in the Carson Sink north of Fallon and one located in the Clan Alpine Mountains east of Fallon. Abe Fortas, acting secretary of the interior, granted the Navy a permit to use the ranges. He made it subject to the condition that livestock grazing would be allowed on a part-time basis. The permit noted that the lands would remain open to mining since the range was being authorized by permit, rather than withdrawal. The permit was to expire no later than six months after the end of the national emergency declared for World War II.[28]

Four months after it received the permit, the Navy objected to the periodic use by livestock.

Because of overcrowded and limited coastal facilities, the Naval Aerial Gunnery program presents a most acute problem. The intensity of that program together with the overcrowded conditions in that area will not permit periodic interruption in these activities. Accordingly . . . it is requested that the Secretary of the Interior make the land in question available to the Navy Department unconditionally in order that uninterrupted year round use may be made thereof.[29]

Livestock grazing was subsequently eliminated from the bombing range. Further expansions of the Fallon training complex occurred in 1945 with a permit for a new bombing range, Bravo 19, located in the Blow Sand Mountains south of Fallon. The permits for the ranges remained in effect until 1952, after President Truman had declared an end to the World War II national emergency.

Under authority delegated by President Truman, in 1953 the secretary of the interior signed a public land order to officially withdraw three of the bombing ranges that exist today: Bravo 16 in Lahontan Valley, Bravo 17 in Fairview Valley, and Bravo 19 in the Blow Sand Mountains. The total size of the withdrawal was 56,000 acres.

The Navy also applied for a fourth range, Bravo 20. However, due to a misunderstanding about the length of the original permit, the request was withdrawn. The Navy operated under the assumption that it was leasing the range from the Truckee Carson Irrigation District. The lands in question, 20,000 acres in the Carson Sink, were under a withdrawal to the Bureau of Reclamation for the Newlands Reclamation Project. The irrigation district operated project facilities for the bureau. The irrigation district was empowered to lease lands only for reclamation purposes. Nevertheless, the Navy continued to use these lands as a bombing range for thirty years without official authorization.[30]

The Bravo 20 range reemerged as an issue in 1974 as the result of an application to the Bureau of Land Management for oil and gas leases on the site. The application prompted a record search that revealed that Bravo 20 was never properly withdrawn. The bureau told the Navy that a congressional withdrawal as required by the Engle Act of 1958 would be necessary for continued use of the range. The Navy was warned that the land was not withdrawn from public access or mining and that claims could be filed until it submitted a withdrawal request. After an extensive debate by memoranda, the Navy agreed and submitted a withdrawal request. That request had the effect of closing access to the area for two years.

In 1976, Congress passed the Federal Land Policy and Management Act (FLPMA). It required a much more open planning process for military withdrawals than the Engle Act. Section 204 of FLPMA required the following information:

1. An inventory and evaluation of the current natural resource uses and values of the site and adjacent land and how
it appears they will be affected by the proposed use.

2. An economic analysis of existing uses.

3. An analysis of the manner in which the land will be used.

4. A statement as to whether any suitable alternate sites are
available.

5. A statement about public consultation.

6. A statement about the effect on state and local government and the regional economy.

7. The place where the applicant's records on the withdrawal
can be examined by the public.

8. A report prepared by a qualified mining engineer on mineral production, mining claims, mineral leases, evaluation of
future mineral potential, and market demands.

To meet these requirements, the Navy held a public hearing,
conducted a mineral inventory, and prepared an environmental
assessment that included the required information. It was submitted to the BLM in 1981 after the temporary closure had expired.
The expiration caused a disagreement between the Navy and the
BLM over whether FLPMA's more stringent requirements were
valid in cases involving the Engle Act.

> The [BLM] Solicitor's opinion is not an objective, impar-
> tial look at the law but that of an advocate, for a principle,
> with a vested interest. Historically, segregation under the
> Engle Act continued from the date of filing the application
> until Congress acted to approve or deny the withdrawal.
> Please note that Congress specifically acted to leave the Engle
> Act in effect when passing FLPMA. Therefore, the termina-
> tion of segregation . . . is inappropriate. . . . You are hereby
> advised that the Navy will deny entry upon the land to any
> party for reasons of safety and security.[31]

Conflicts between agencies such as the BLM and the Navy
are not necessarily counterproductive. They can be viewed as democracy in action—reflecting a constitutional system that pro-

vides for different agency missions and divides and separates power. Nevertheless, the Navy maintained control over Bravo 20 while it completed its withdrawal requirements. Legislation was submitted to Congress for the Bravo 20 withdrawal with a recommendation that it be approved as proposed by the Navy.

The withdrawal bill had great difficulty in Congress. Most of the debate over the withdrawal occurred in the House of Representatives and focused on the withdrawal period. Republicans pushed for a twenty-five-year withdrawal while Democrats preferred a shorter ten-year period. The debate also involved the adequacy of the environmental analysis for the withdrawal. Republicans contended that the existing assessment was adequate while Democrats pushed for more detailed analysis and requested a comprehensive environmental impact statement on all military activities in Nevada.

Major differences on the scope of the withdrawal bill were also debated. Republicans wanted to pass the Bravo 20 proposal as a separate bill, while Democrats wanted to pass a more comprehensive bill involving other military withdrawals in Nevada and the western states.

The debate extended through 1985. In June of that year a group of protestors led by rural Nevada activist Dick Holmes challenged the Navy's right to close the range by occupying Bravo 20. They claimed that it was open public land illegally closed by the Navy. This protest and subsequent mining claims filed by the group forced the Navy to cease its use of the range, initiate a bomb cleanup operation, and reapply for a legal withdrawal. Congress finally approved the withdrawal of Bravo 20 in the omnibus Military Lands Withdrawal Act of 1986. That act is analyzed in Chapter 8.

The Fallon Naval Air Station is currently the Navy's premier aircraft carrier strike force and bomber training facility. It is the largest employer in Churchill County, accounting for about 2,000 jobs. The population associated with the station amounts to about 6,500 people or 41 percent of the county's 16,000 residents.[32] Many of the other residents rely on farming for their livelihood. Fallon is the site of the Newlands project, the nation's first reclamation project. Since 1907 it has provided water for irrigating thousands of acres around Fallon. To the east, scattered livestock ranches winter their cattle on Indian ricegrass and other forage that grows among the shadscale and greasewood shrubs in Dixie and Edwards Creek valleys. In the summer, cattle feed on high-country meadowlands in the Stillwater, Clan Alpine, and Desatoya mountains.

After a tour of duty at sea, a carrier air group is dispersed to various home bases. In the west, fighter (F-14) wings return to the Miramar or Top Gun Air Station in California, light attack (F/A-18) wings return to Lemoore Air Station in California, and medium attack (A-6) wings return to Whidbey Island Air Station in Washington. Before they are redeployed at sea, carrier air groups assemble for joint training at Fallon. These involve up to 1,200 personnel and seventy-five aircraft for two weeks of intensive training on all the Navy's Bravo bombing ranges.

Bravo 16 is southwest of the air station. Like the other Bravo ranges, it has a remote video scoring system. The Navy uses it for nuclear bomb training with inert ordnance. No nuclear bombs are dropped.

Bravo 17, east of the air station, is the most heavily used range. It is used for strafing, dive bombing, and rockets using both live and inert ordnance. The Navy uses it with electronic warfare sites that jam radar and simulate enemy threats such as surface-to-air missiles. It has targets similar to the Nellis Range. These include a mock army camp, industrial site, power plant, enemy runway, weapons revetments, and dummy missile sites with electronic emitters. Most of the targets were built in the late eighties. The Navy recently installed a no-drop bomb scoring system to further enhance Bravo 17's capabilities.

Bravo 19 is used for strafing practice and live bombing sorties. It is located south of the air station at Rawhide Flat. Pilots practice close air support and strikes on truck convoys.

Bravo 20, in the Carson Sink, is used for air-to-ship bombing practice with the Navy's heaviest bombs. Pilots also use the range for dumping ordnance when they have trouble with their aircraft.

The Electronic Warfare or "Echo Whiskey" Range is just north of Bravo 17 in Fairview Valley. It was created during the Vietnam War and has the capability of simulating threats from Soviet, Iraqi, or other enemy surface-to-air missiles and radar.

Unlike the other ranges, it is not withdrawn for exclusive Navy use. Electronic warfare sites are permitted by individual rights-of-way. No bombing is permitted. Livestock grazing, recreation, and other public land uses are permitted.

Much of the current activity at the air station can be attributed to events in Lebanon. On December 4, 1983, a Navy strike force attacked Syrian ammunition dumps, radars, and surface-to-air missile sites in Lebanon. The strike was a disaster, with two aircraft lost and one pilot killed. The Lebanon debacle prompted the Navy to intensify its strike warfare training, the most significant changes occurring at the Fallon Naval Air Station.

Recent additions at Fallon have included the establishment of a Supersonic Operations Area that allows the Navy to operate its newer aircraft at supersonic speeds. The Navy also established an adversary squadron to provide realistic air-to-air threats similar to the Air Force's Red Flag operations.

The most important of the new additions at Fallon was the establishment of a Strike Warfare University that provides training for the planning and execution of carrier battle group bombing tactics. It was modeled on the Navy's "Top Gun" fighter training school at the Miramar Naval Air Station in California. Part of the Strike Warfare University's function has been to refine existing concepts like the Alpha Strike to overwhelm air defenses with massive numbers of planes striking simultaneously. It is also charged with developing innovative new low-altitude tactics to counter sophisticated air defense systems that even third world nations such as Iraq have obtained.

Another recent addition to the Fallon Naval Air Station was the electronic Tactical Aircrew Combat Training System similar to the one at Nellis. It tracks and records aircraft altitude, speed, and distance from other aircraft. It can determine scoring accuracy and pilot responses to both air-to-air and surface-to-air threats.[33]

The Nevada Nuclear Test Site

Although established later, the Nevada Nuclear Test Site had its roots in World War II. Following the end of the war, the United States began a nuclear testing program in the Marshall Islands. Logistical problems resulted in the initiation of a top secret project,

code-named "Nutmeg," to find a suitable site in the continental
United States. The project report concluded that the decision to
establish a site would depend "upon the elements of public relations, public opinion, logistics, and security." It was felt that the
establishment of testing within the United States would require a
national emergency.[34]

That emergency began in the fall of 1949 when the Soviet
Union exploded "Joe One," its first atomic bomb. The emergency
deepened considerably in the summer of 1950 when the United
States became involved in the Korean War. By December 1950,
President Truman had approved a site on the Nellis Bombing and
Gunnery Range. Six weeks later, the first atomic weapon was
tested over Frenchman Flat.

The planning process for the establishment of the continental
site occurred in total secrecy and with careful calculation of public
reaction. Atomic Energy Commission planners originally studied
five sites: White Sands, New Mexico; Dugway, Utah; Camp Le-
jeune, North Carolina; the Shoal Site east of Fallon; and Nellis in
southern Nevada. They selected Nellis because it was already
under federal ownership; it included Camp Mercury, a temporary
air base that could be easily converted to an operations center; it
had the support of Nevada Senator McCarran; Nevada residents
were expected to welcome its economic contributions; its isolation
assured security problems would be minimal and accidents would
affect only a few people; and the dry climate and predictable winds
were advantageous for atmospheric testing.[35]

Truman issued a very short public announcement that ex-
plained only that experimental nuclear detonations would take
place. Public comments were not requested. Convincing Nevadans
of the need for continental testing was easy following the Soviet
explosion of Joe One. The military capitalized on the public concern generated by the Soviet threat to establish the site and boost
appropriations for weapons development and testing. An Atomic
Energy Commission spokesman emphasized the dangers of letting
America's guard down.

Soviet Russia possesses atomic weapons, there is no monop-
oly for the free world. Therefore, we have no alternative but

to maintain our scientific and technological progress and keep our strength at peak level. The consequences of any other course would imperil our liberty, even our existence.[36]

Legal transfer of the test site to the Atomic Energy Commission (now Department of Energy) from the Air Force was mandated by Public Land Order 805 in 1952. It amounted to 435,000 acres. An additional 318,000 acres were added in 1961.

During the early atmospheric testing, thousands of southern Nevada and southern Utah residents were issued badges to monitor radiation exposure. At first there wasn't any serious opposition to the testing, but after several years of monitoring it became apparent that the downwind impact of fallout might be greater than expected. State leaders, including Governor Charles Russell, felt powerless to do anything about it:

> The federal government did not request permission to use the public domain in Nevada for the atom bomb tests. You must realize that the area set aside by the government is and always was, federal land as is more than 86 percent of the entire area of Nevada. Since all of the atomic tests in Nevada are carried out upon federally-owned property, there is nothing that I as Governor . . . can do to halt such tests.[37]

The Nuclear Test Ban Treaty of 1963 finally ended aboveground testing. The controversy over health effects of aboveground testing has continued ever since and is currently the subject of many lawsuits. A side effect of the controversy was that many Nevadans began to question whether the military establishment in Washington had their best interests in mind.[38]

FOUR

The Engle Act

ongress became increasingly concerned about military expansion in Nevada and elsewhere during the fifties. In 1958 it passed the most significant legislation to date regarding military land use planning activities. Public Law 85–337 is commonly known as the Engle Act. The act requires that congressional approval must be obtained for any military withdrawals over 5,000 acres.

During an eighteen-month period between January 1955 and June 1956, the various military services had applied for more than 14 million acres of public land, including 3 million acres in Nevada alone. According to California Representative Clair Engle, congressional action was necessary because the military had become "awful land hogs."

> What they were seeking would have amounted to nearly 26,000 acres a day. To be precise—1,000 acres an hour. To you fellows from the city, that would figure out to a military public land demand of 17.7 acres per minute over a period of nearly 800,000 consecutive minutes.[1]

The Department of the Interior systematically approved applications for public land by the military services. Presidents Roosevelt and Truman had delegated the broad implied withdrawal authority of the executive branch to the secretary of the interior by executive orders in 1942, 1943, and 1952. These executive orders gave the secretary the power to issue withdrawals under public land orders.

Nevada Senator Alan Bible, during debates over a prototype of the Engle Act, stated that the Department of the Interior had not been able to compare the value of existing public land uses with the need for national security as identified by the military. Consequently, requests for additional military lands were routinely approved.

During testimony at the House hearing on the bill, Department of Defense officials admitted that the military had "never had a turndown from the Department of the Interior."[2]

The Army, the Navy and the Air Force would simply make out a slip of paper in the nature of an application to the Interior Department asking for an area perhaps 100 miles long and 50 miles wide in the state of Nevada, and send it over to the Secretary of Interior saying it was absolutely necessary to their operations, and that area was set aside for those military operations and put into what could be regarded as a legal icebox. . . . It was [congressional] concern over that problem that brought on this legislation.[3]

During this period most of the lands in question were considered desert wasteland, worthless except for occasional hunting and fishing. Shortly after the Engle Act debates, the Wilderness Act omitted the public domain lands managed by the Bureau of Land Management. This was not the outcome of extensive debate or a reflection of public land planning policy, but merely an oversight. "Four-hundred and sixty-five million acres of public domain lands were simply forgotten."[4]

Congressional concern over military land withdrawals probably would not have been so extensive had it not been for the concern of hunters over military restrictions to hunting areas. The armed services generally denied public hunting and fishing on military reservations. Exceptions were made for members of Congress, city officials of adjacent communities, retired military personnel, invited guests of military personnel, and prominent citizens who "have demonstrated active interest in military affairs."[5] Given the nature of these groups, Congress questioned the military's assertion that closure to the general public was necessary for security reasons. Still, it was difficult to refute the military's arguments. Most of the bases operated under a level of secrecy that prevented public knowledge of real security needs. Representative John Saylor of Pennsylvania drew his own conclusions.

Most of these reservations, the testimony to the contrary notwithstanding, are private game reserves for the military. Some of us who are associated with the military know. We have been there and seen them.[6]

Congressional pressure may have led to the opening of some military reservations for limited hunting and fishing access, including the Navy's Hawthorne Ammunition Depot. Yet Congress did not feel that "jawboning" was sufficient. Open public debate before its subcommittee for each new withdrawal was necessary.

A key issue in the debate was constitutional authority to manage public lands. Over the years by practice and through various legislative actions, Congress had been relieved of its constitutional responsibility to manage these lands. Article IV, Section 3, Clause 2 of the Constitution declares:

The Congress shall have power to dispose of and make all needful rules and regulations respecting the Territory or other property belonging to the United States.

The Engle Act returned this responsibility to the legislative branch for federal military withdrawals over 5,000 acres by requiring an act of Congress for any such proposal. The idea was that Congress would set the terms and conditions of each specific withdrawal after seeking the advice and assistance of the public.

Procedures to carry out the Engle Act were worked out by the Departments of Interior and Defense. They include the submission of an application for withdrawal by the military to Interior's Bureau of Land Management.

An application causes the proposed lands to be protected from appropriation by miners under the 1872 Mining Law for a two-year period during which the withdrawal is processed. After an analysis of the lands to be withdrawn, the secretary of the interior submits legislation to Congress for the military proposal. The secretary may also submit Interior's own recommendations, if they differ from the military's.

This process was changed substantially with the passage of the National Environmental Policy Act in 1969. That act states that it is national policy to prevent or eliminate damage to the environment and biosphere. It includes policies for fulfilling the federal government's responsibility as trustee of the environment for succeeding generations, preserving the natural aspects of our national heritage, and maintaining an environment which supports diversity and variety of individual choice. This is to be accomplished primarily through the preparation of environmental impact statements for federal actions that could have a significant impact on the environment.[7]

The courts have decided that the National Environmental Policy Act is a procedural law. It requires federal agencies to follow an established procedure in preparing environmental impact statements. It does not require them to select an environmentally preferable course of action.

Engle Act withdrawals usually cause significant adverse environmental impacts and therefore require the preparation of environmental impact statements. The EIS process allows citizens to participate in the review of military withdrawal proposals. Agencies conduct public scoping to decide what issues should be analyzed in a withdrawal EIS. Citizens can submit written comments and participate in public hearings on draft EISs. They have the right to sue the federal government for not complying with National Environmental Policy Act procedures and may testify before congressional committees. From a public perspective, the EIS process is the planning process. Its major weakness from a planning viewpoint is the lack of a requirement for up-front public involvement in the development of proposals.

Except for some boundary adjustments to the Nellis Range, military planners left Nevada alone after the passage of the Engle Act in 1958 until the mid-seventies. Then, air space as well as land withdrawals became a major issue. Although the Engle Act itself did not address military air space demands, the House Interior Committee recognized that they would eventually require attention:

It is clear to this committee, then, that military use requirements today must be thought of in terms of both horizontal and vertical needs. While the concern and jurisdiction of this committee is limited to the former . . . the committee believes that it is absolutely vital that continuing reevaluations be made of Federal legislation and administrative controls governing the assignment and use of air space, which does involve the latter.[8]

Air space conflicts first became a major concern in Nevada during the debate over the Continental Operations Range. That debate continues today.

FIVE

Air Force Proposals

n the seventies and early eighties, a series of
Air Force projects generated long-term reservations about the military's role in Nevada.
Unacceptable pilot and aircraft losses during the early phases
of the Vietnam War set the stage for renewed military interest in
the deserts of Nevada. The Air Force focused much of the blame
for those losses on inadequate pilot training. Lessons learned in
World War II and Korea were not working because of advances in
aircraft and weapons technology. The Air Force decided that it
needed more realistic training.

In early 1966, the Air Force developed general concepts for
this effort, which included the controversial Continental Operations Range (COR). After internal study and evaluation, the secretary of defense approved a proposal for the COR to be developed
in the west-central United States. By 1972, the Air Force's Tactical
Air Command narrowed the search to Utah and Nevada.

During the planning process for the Continental Operations
Range, the Air Force decided that its training exercises needed
two-sided engagements for air-to-air combat and realistic ground-
based threats such as surface-to-air missiles.

Air Force planners concluded that existing facilities had three
major weaknesses: civilian "encroachment" on ground, water,
and air space; saturation of existing capacity; and weather limitations. They decided that existing ranges were inadequate because
they could not provide training conditions that represented enemy
ground and airborne air defense systems. They did not have

enough air space for air strike force tactics and could not accommodate supersonic operations.

The Air Force studied two simulator-based alternatives. One type involved "no-drop" bomb scoring in real aircraft. It would eliminate the need for land, but not air space. The other type consisted of total aircraft simulators that would eliminate the need for both air space and land. The Air Force decided these were unacceptable due to lack of realism. Its conclusion was that only a continental operations range would be adequate.[1]

Planning criteria developed for the COR were:

1. Provide a range facility which would permit Operational Testing & Evaluation of equipment for strike-sized forces (one to 100 aircraft) in a realistic combat environment.

2. Provide large land and air space areas where exercises could be conducted with a minimum of constraints to train military air warfare elements in a realistic but simulated combat environment, and to evaluate tactics, performance and capabilities of those elements.[2]

The Air Force decided that only one area could meet its needs. This was a triangular region including its Nellis Range near Las Vegas and the Utah Training Range near Salt Lake City and the Navy's Bravo Ranges near Fallon (Map 4). They selected the Nevada/Utah site "on the basis of [low] air traffic density." Other factors included: the high percentage of federally owned land would result in minor land acquisition costs; mild weather conditions were conducive to training activities; the area had a wide range of topography; its inland location protected it from surveillance from unfriendly submarines; it had a low population density; and it incorporated existing facilities.

The Continental Operations Range consisted primarily of air space reservations rather than land withdrawals. It involved expansion of air space reservations at Nellis and establishment of air routes between Nellis, the Utah Training Range, and Fallon. It included more land withdrawals east of the Nellis Range in Lincoln County for electronic warfare sites. These would consist of communications equipment that would electronically emit threats to training aircraft such as simulated surface-to-air missiles.

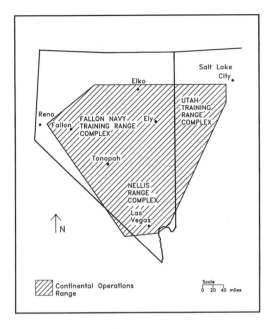

Map 4. Continental Operations Range

The COR was intended to allow for integrated Navy/Air Force operations. It called for electronic exercises, actual flights, and simulated attacks between the bases as well as live bombing runs. The program was expected to cost $300 million and provide 700 new jobs.[3]

The Air Force believed that the Continental Operations Range project would increase air safety because of better air traffic control. The Air Force anticipated some inconvenience to private pilots from increased air space reservations. It was also concerned about electromagnetic emanations from the electronic warfare equipment, noting that there existed "a remote possibility of errors in operations procedures whereby some nonparticipants (as well as participants) could be illuminated by main beam radiations."[4] The Air Force also expected adverse impacts to people and animals from increased sonic booms and low-level subsonic flights. On the other hand, it projected significant beneficial economic impacts, particularly in Tonopah and Caliente. Based on this analysis, the Air Force concluded that the project was noncontroversial.

During the early stages of the development of the proposed
COR . . . rumors were circulated and some private pilots in
the area expressed objections to the project as they then perceived it. The Air Force subsequently made exhaustive efforts
to inform the public and correct any misconceptions regarding
the proposed COR. . . . Apparently due to this extensive program, as of this writing the potential controversy with the
private pilots appears resolved.[5]

At first the public relations effort was successful and the
project received public support. Senator Wallace Bennett, R-Utah,
was a supporter of the project due to the potential for increased
military and civilian employment in Utah.[6] The people of Tonopah,
who anticipated an additional fifty jobs in their community, also
welcomed the COR.[7] But with the end of the Vietnam War, Congress balked at the project's price tag and the Air Force scuttled it
in 1975.[8]

Despite its early demise, the Continental Operations Range
has surfaced time and again in military planning efforts in Nevada.
Opponents of military expansion in the state have accused various
military departments of "piecemealing" the COR through many
smaller projects.

The MX Missile

Air Force planners next focused on the Great Basin of Nevada
and Utah in 1977, this time not for training pilots, but for basing
intercontinental ballistic missiles. The Air Force had concluded
that fourth-generation Soviet missiles had rendered the land-based
U.S. missile system vulnerable to a first strike attack. The only
way to counter the threat was with the MX (for missile-experimen-
tal). The Air Force maintained that if MX missiles were not operational by the early eighties, an unacceptable risk to the nation's
strategic forces would exist.[9]

The planning process began with consideration of the entire
continental United States. Application of planning criteria quickly
narrowed the search to seven sites, all in the West. The criteria
involved primarily geological and physical factors, but also the
presence of population centers, parks, and Indian reservations.

In June 1979, President Carter announced that the MX would be based in 4,600 launching sites scattered across the deserts of Nevada and Utah. This decision involved additional planning criteria, including a minimum distance of 200 miles from the coast and international borders and compatibility with local land uses. Air Force planners wanted to maximize use of federally owned land and focus on areas with very low rural populations, low economic activity levels and ample undeveloped land.[10]

The project consisted of 200 mobile MX missiles, each missile having twenty-three shelters. Missiles would be moved on an oval track at random among the shelters in order to confuse the Soviets. This was the racetrack, or Multiple Protective Shelter (MPS), basing mode. The system would be dispersed over an area of 5.4 million acres and create 160,000 acres of surface disturbance. Air Force planners proposed a point security system to minimize the amount of land that would be excluded from public use. This involved withdrawing and fencing the shelters only, all other land being open for livestock grazing, mining, and recreation.

The public first took a wait-and-see attitude about the proposal. Business interests in Las Vegas were early supporters. They expected to benefit from huge government construction contracts. The MX was projected to cost as much as $56 billion, the most expensive project in U.S. history.[11] Public opinion polls revealed that 60 percent of the state's residents initially supported the project.[12]

Public opposition grew rapidly after the release of a 1,900-page draft environmental impact statement in December 1980. The MX proposal came on the heels of the "Sagebrush Rebellion," a states-rights movement promoted by the livestock and mining industries. The rebels opposed the environmental protection programs of federal land managing agencies. Backed by the state government in Carson City, the rebels tried unsuccessfully to have federal public lands given to the state and sold to private developers. A coalition of environmentalists, Native Americans, the Mormon Church, and others were joined by the Sagebrush Rebels in a "Sagebrush Alliance" to oppose the MX. Intense emotions were generated.

Here, not among the cattlemen of the north or the politicians of Carson City, was the genuine sagebrush rebellion, a grass-roots reaction to a military gone mad.[13]

Hostility to the project was generated by several related concerns. Land use issues were significant. They included the potential loss of land for grazing, mining, and recreation. Portions of the area had been proposed for wilderness and for the Great Basin National Park. Other issues included: the project's use of the region's scarce water resources; the potential for the area to become a nuclear target; and disruption of the native ecosystem and loss of wildlife habitat. But the most volatile issue involved the population influx that would be generated by the MX. Demands for housing, schools, and other public services were expected to skyrocket, along with taxes to pay for them. Also significant was the effect that rapid growth would have on the peaceful rural life-style of the region.[14]

Governors of both states reacted strongly after it became clear that the public opposed the proposal. Governor List of Nevada demanded that Congress either give or sell public lands to the state if the MX was approved. Utah's Governor Matheson took a different approach. He insisted that the Air Force had to make the "strongest case possible" that the MX proposal was necessary for national security and that it was being done in the least damaging way.[15]

During his 1980 presidential campaign, Ronald Reagan declared himself a "Sagebrush Rebel" and expressed major reservations about the proposed basing mode for the MX. Congress also had reservations about the proposal. They questioned the point security system in particular. A House of Representatives report concluded that the Air Force would be unable to guarantee the security of the MX. It would eventually have to eliminate public access to far more acreage than it was proposing under the point security system. This would mean that the social and economic impacts of the MX would be far greater than projected.[16]

The House also believed that the proposed MX Multiple Protective Shelter system would provide only the illusion of survivability and could escalate the arms race. Its conclusion was to scrap the proposal.

Generally, the deficiencies of the proposed MPS basing system fall into three categories: strategic defects, social and economic problems and environmental problems. While the second and third categories are of great importance, they are necessarily secondary to the strategic concerns of national defense. But the strategic defects of MPS are so grave that the national security purposes of the proposed basing scheme cannot justify the expense—the land, money, human skills, time, and other resources required—to deploy the MX missile in such a basing mode considering that there are better alternatives.[17]

MX deployment in the Great Basin was eventually scrapped. It is currently being deployed in hardened silos in the Midwest. The MX has had a major impact on subsequent military land use planning efforts in Nevada. Nevadans learned that by putting their other differences aside and cooperating on a single issue, they could have a major effect on decisions made about their lands in Washington, D.C.

The Groom Range

Another controversial Air Force project involved an attempt to avoid the Engle Act by direct appropriation of land. The Groom Range is a north-south–trending mountain range in Lincoln County next to the Nellis Range. It is flanked by broad alluvial valleys. Bald Mountain at 9,400 feet is its highest peak. The highest elevations are snow capped in winter and have scattered stands of white fir and limber pine. Lower elevations have extensive stands of sagebrush and saltbush.

The Groom Range is located in Lincoln County in a remote area of southeastern Nevada. Lincoln County is one of the largest (6.8 million acres) and least populated (3,500 people) counties in the nation. Like much of the county, the Groom Range was open public land. It was used mainly for mining, livestock grazing, and hunting, having a reputation for producing trophy-quality mule deer.[18]

The Air Force's interest in the range was its location next to Area 51 of the Nellis Range where it was conducting tests of the stealth bomber and Strategic Defense Initiative defense systems.[19] In 1978 the Air Force began controlling entry to the Groom Range area with armed troops in order to provide a security buffer for projects at Area 51.[20] The closure became common knowledge when the Air Force established security checkpoints in 1984.

Public concern over this action prompted the Air Force to request a congressional emergency closure of the area for national security reasons. At the House of Representatives hearing on the proposal, the Air Force admitted that it "had no legal authority" to control access to public lands; nevertheless, it "asserted the right" to deny access.[21] Nevada Representative Harry Reid said that "closing of the lands under the banner of national security is an arrogant display of power." Governor Richard Bryan saw the action as part of a broader issue.

> For years, Nevadans have acquiesced to defense-related land withdrawals, but the time has come to draw the line. . . . I strongly suggest to you that the day is past when the federal government can look at Nevada . . . as an unpopulated waste-land to be cordoned off for whatever national purpose seems to require it.[22]

Although it was a foregone conclusion that the Air Force would be granted the withdrawal, it was attacked for the way it seized the area. Congressman John Sieberling compared it with Watergate, remembering the illegal actions that had been justified in the name of national security in those days. He believed that Watergate proved that "no man, no institution is above the law in this country."[23] In Nevada, environmentalists questioned the Air Force's priorities:

> The only thing you seem truly concerned about . . . is national security. Yet you completely fail to define the criteria for such needs. It is therefore impossible for a person to evaluate the . . . actions in terms of need. . . . We are all Americans and we are all concerned with national security. However, we cannot put national security up on a pedestal and say that everything else must bow before it.[24]

The Air Force countered by pointing out that, at least after the fact, it had complied with the Engle Act and the National Environmental Policy Act. It allowed the public to participate in the withdrawal process by attending hearings and submitting comments on the Groom Range environmental impact statement. The Air Force considered this a good example of democracy in action. At the EIS hearings in Alamo, Nevada, an Air Force spokesman told the public:

> You are participating in what I consider to be a uniquely American experience, and that is, to have your government tell you in advance of a proposed action it intends to take, furnish you with the details of that action, solicit your comments and points of view on the impact of that action on you and your community, and then take into serious consideration your comments before a final decision is made. And this is done without any threat of reprisal or harassment should your point of view not be agreed with.[25]

It could be argued that the action was taken in advance of public notice—the Air Force was already in possession of the Groom Range. A major problem with the process outlined by the Air Force was the lack of public involvement at the beginning of the planning process, before it decided to appropriate the Groom Range. Nevertheless, public input did have an impact on the final decision. The Air Force accepted proposals to mitigate the impact of the withdrawal by improving off-range wildlife habitat. It decided to open areas of the Nellis Range for hunting and to improve access roads.

On June 17, 1988, Congress passed Public Law 100–338, which legally withdrew 89,000 acres of public land in Lincoln County as a security buffer zone for the Air Force operations on the Nellis Range.

State Military Maneuvers

hile state of Nevada officials maintained a
general opposition to federal military expansion, they had different standards for
nonfederal proposals. State officials backed controversial proposals for private bombing and rocket and state National Guard tank
ranges.

The Aerojet Land Swap

On June 8, 1926, Congress passed an act approving an exchange of lands between the state of Nevada and the Department
of the Interior. It provided that up to 30,000 acres of land of equal
value could be exchanged. Its purpose was to provide a more efficient land ownership pattern. The exchange occurred on a piecemeal basis over a period of years until by 1962 only 2,000 acres
remained to be selected by Nevada.

In response to a proposal by Aerojet-General, a Department
of Defense contractor, Nevada applied for 1,130 of its remaining
acres. The land was located in Garfield Flat, a remote dry lake
valley in Mineral County. The flat itself was barren of vegetation.
Surrounding lands supported saltbush stands. They were occasionally grazed by livestock and wild horses. Soon after obtaining the
land, the state sold Garfield Flat to Aerojet.

The defense contractor used the lands for testing ammunition
and bombs. Garfield Flat essentially amounted to a Department of
Defense bombing range, although privately operated by one of its
contractors.

For safety reasons, Aerojet began applying for various buffer zone permits to close adjacent public lands during bombing missions. The Bureau of Land Management granted those permits routinely until 1986. By that time the bureau had received a legal interpretation of the Federal Land Policy and Management Act that concluded that it lacked authority to grant permits for buffer zones for military purposes without a withdrawal. It therefore denied an application by Aerojet to extend its buffer zone permits.[1]

In response to the military buildup of the eighties, Aerojet had been searching for an area to build and test rocket engines for the Department of Defense. By 1986, it had narrowed its choices to one location, a 50,000-acre site in Coyote Springs Valley on the Lincoln/Clark County border in southern Nevada. The site was attractive because it was remote from population centers and had a productive well that had been drilled during the analysis of the MX missile project. The well was needed to provide water for Aerojet's rocket engine production program.

Aerojet had purchased about 60,000 acres in Dade County, Florida, in the fifties. It planned to use the property to test liquid rocket fuel. The project failed when rocket development focused on solid fuels. In 1983, Aerojet made a bargain sale to the Trust for Public Land of 50,000 of its acres. The trust resold the land to public agencies.

Aerojet leased about 4,500 acres of its remaining lands, which were located next to Everglades National Park, for tomato farming. The South Florida Water Management District desired this land for water diversion purposes and to protect the park.[2]

Aerojet, after initial feelers to the National Park Service about this land, found a willing partner in the U.S. Fish and Wildlife Service. Together, they developed a scheme to trade public lands administered by the Bureau of Land Management in Nevada for Aerojet's lands in Florida. The Nevada lands would be used by Aerojet for a buffer zone around its Garfield Flat bombing range and to establish a rocket production and testing range at Coyote Springs Valley for Department of Defense projects. Aerojet's Florida tomato fields would be sold to the South Florida Water Management District. The proceeds would be used to fund U.S. Fish and Wildlife Service land acquisition strategies in Florida.[3]

Southern boundary, Hawthorne Army Ammunition Depot, Mineral County, Nevada. (David Loomis collection)

Aerojet General's Garfield Flat Bombing Range, Mineral County, Nevada. (David Loomis collection)

Entrance to Cottonwood Canyon, Mount Grant area,
Hawthorne Army Ammunition Depot, Mineral County,
Nevada. (David Loomis collection)

Ammunition facilities, Mount Grant in background,
Hawthorne Army Ammunition Depot, Mineral County,
Nevada. (David Loomis collection)

Ammunition bunker, Hawthorne Army Ammunition Depot, Mineral County, Nevada. (David Loomis collection)

Army hunting cabin, Mount Grant area, Hawthorne Army Ammunition Depot, Mineral County, Nevada. (David Loomis collection)

Cottonwood Canyon, Mount Grant area, Hawthorne Army Ammunition Depot, Mineral County, Nevada. (David Loomis collection)

Main entrance, Fallon Naval Air Station, Churchill County, Nevada. (David Loomis collection)

Naval electronic warfare site, Sand Springs Range, Churchill County, Nevada. (David Loomis collection)

Naval bomb storage bins, Bravo 19 Range, Churchill County, Nevada. (David Loomis collection)

Navy target vehicles, Bravo 17 Range, Churchill County, Nevada. (David Loomis collection)

Navy target bus, Shoal Site, Churchill County, Nevada. (David Loomis collection)

Bravo 19 Range, Churchill County, Nevada. (Warren Loomis collection)

Mock industrial facility, No-Drop Bombing Site, Bravo 17 Range, Churchill County, Nevada. (Warren Loomis collection)

Navy bomb-scoring tower, Bravo 16 Range, Churchill County, Nevada. (Warren Loomis collection)

Navy F/A-18 "Hornet" fighter/bomber on the tarmac at Fallon Naval Air Station, Churchill County, Nevada. (Photo by Randy Duran)

Defense contractors had promoted similar land swaps in Nevada in the past. In the fifties, North American Rockwell Corporation obtained public lands northeast of Reno for a rocket testing facility. Although North American used the facility for a short period, soon after the corporation obtained clear title, it sold the land for scattered subdivisions. Contaminated wells are a problem for current landowners.

In 1957, another Department of Defense contractor, Curtiss-Wright, developed a land swap proposal to acquire most of the public land in Storey County for a 140,000-acre rocket research and testing center. The New Jersey–based corporation swapped land it had purchased from Southern Pacific Railroad for 78,000 acres of public land in the Virginia Range southeast of Reno. It added a 60,000-acre ranch along the Truckee River for its proposed rocket center.

But Curtiss-Wright failed to win its rocket contract, and the research center never got off the ground. "Chamber of commerce dreams" for 10,000 new jobs faded with it. Despite selling some of its land to subdividers over the years, Curtiss-Wright still owns most of Storey County. It receives an agricultural deferment for property taxes as marginal grazing land. Its taxes amount to only $3,000 per year.[4]

In 1987, the company sold 11 square miles of its land swap land to Hi-Shear, a California-based Department of Defense contractor. Hi-Shear is currently under investigation by the Nevada Department of Environmental Protection for hazardous waste violations on the property.[5]

Apparently, for the Aerojet proposal, the history of land swaps in Nevada and real estate deals in Florida were overlooked. Nevada politicians saw the Aerojet proposal as an opportunity to diversify the state's economy. Governor Bryan, the state legislature, and Nevada's congressional delegation came out in early support of the swap. They were attracted by Aerojet's initial suggestion that it might create 2,000 jobs in Nevada. This estimate was later revised to about 600 jobs.[6]

Potential tax benefits were also used to generate support. Representative Barbara Vucanovich maintained that the swap was vital because it would get more land on the tax rolls.[7] Supporters believed this was particularly significant in Lincoln County, where most of the land is in public ownership. Public lands are not subject to local property taxation. Local governments are only partially compensated through the federal Payments in Lieu of Taxes program.

Aerojet proposed to build its Coyote Springs project in three phases. Each would involve construction of eighteen buildings covering 2.5 acres. The buildings were to be used to store chemicals and house rocket engine assembly and testing facilities. Aerojet projected a total land disturbance of 300 acres.

Aerojet wanted the remaining 49,700 acres for a buffer zone for public safety and for security reasons. The land was to be posted with "No Trespassing" signs and patrolled by security guards to prevent public access. Since the project depended on Department of Defense contracts, it was unknown how long it would be in operation. If and when the project ended the lands were to be sold on the open market.[8]

The corporation did not propose any new employment for its Garfield Flat bombing range expansion. The 8,900 acres from the land swap would be used only as a buffer zone. The only changes from the land swap would be an extension of a fence to further restrict public access to the flat and more "No Trespassing" signs. Existing bombing operations on the range were to continue.

Aerojet's Department of Defense contract calls for bombing Garfield Flat once a month. The contract runs through 1992. The purpose is to ensure that cluster bombs produced by Aerojet in California function correctly. Cluster bombs consist of many grenade-type weapons encased in a larger bomb shell. They are primarily an antipersonnel weapon.

Aerojet has no direct employees in Nevada for its bombing range. It uses workers from the Day, Zimmermann, and Basil Corporation on a part-time basis. That corporation also manages the Army's Hawthorne Ammunition Plant and stores the cluster bombs there. Once a month it takes them to the Fallon Naval Air Station where they are loaded on bombers. The public is cleared from the Garfield Flat area and all roads are blocked until after debris from the bombs is cleaned up. The debris is disposed in Nevada to Aerojet's satisfaction.[9]

Land exchanges *within* a state are legal under the Federal Land Policy and Management Act. Such exchanges occur in an open manner with full citizen involvement. They are subject to the National Environmental Policy Act with its requirements for comprehensive environmental impact statements. The act also provides citizens the right to challenge inadequate environmental analyses in court.

Since the Aerojet exchange involved two states, only Congress could approve it. In such cases, Congress may ask for compliance with environmental laws. It chose not to for the Aerojet deal.

The Nevada congressional delegation attempted to get the exchange through in 1986 without public hearings. The plan fell through after Nevada and Florida environmentalists objected to the lack of citizen involvement. They convinced Ohio Senator Howard Metzenbaum to delay the bill until Congress provided public review.[10]

Environmentalists believed the deal was being rushed through without proper consideration of environmental impacts or public concerns. Given additional time, Nevadans began to focus on Aerojet's past record.

Aerojet had been dumping toxic chemicals at its rocket manufacturing plant near Sacramento since the fifties. That practice was common because little was known about how harmful the chemicals were. In the seventies, regulators found toxic chemicals in wells near Aerojet's property. After a lengthy legal battle with federal and state regulators, Aerojet agreed to pay fines of more than $7 million. It also agreed to begin a program to clean up the groundwater. The program will take several decades and cost more than $10 million.[11]

The U.S. Fish and Wildlife Service also had more time to study the land swap. The service's Great Basin field office concluded that the Coyote Springs part of the project could intrude on critical desert tortoise habitat and a bighorn sheep migration route. More importantly, it could jeopardize the continued existence of four endangered fish: the Moapa dace, Pahranagat roundtail chub, White River springfish, and Hiko White River springfish. The field office biologists based their conclusion mainly on Aerojet's long-range plan to sell the property on the open market when its rocket facility closed. The field office biologists expected that potential farm or subdivision development would result in a major increase in groundwater withdrawals. Since the fish depend on the groundwater aquifers, any decline in water levels could be deadly.

The service's regional headquarters staff in Portland agreed with the conclusion about adverse impacts to the desert tortoise but felt that the analysis of impacts to the endangered fish was too speculative. They overturned the jeopardy opinion, clearing the way for the land swap to proceed. This was despite their conclusion that other releases of public land in Nevada have resulted in the extinction of at least one species and the listing of eleven other species as threatened or endangered.[12]

Meanwhile, in response to public demands, Aerojet paid for an "environmental assessment." The assessment was solely a private report, not subject to National Environmental Policy Act regulations. As such it didn't have objective federal agency analysis procedures, only those required by Aerojet. The report came under severe criticism.

The Preliminary Draft Environmental Assessment prepared by Resource Concepts of Reno for Aerojet is a contaminated document . . . it is totally inadequate and can honestly be described as a "sweetheart" document.[13]

Aerojet does try to make their "environmental report" appear to be an "Environmental Assessment"; however the process in producing their report is fundamentally flawed . . . they did not involve the public at an early stage, and public agencies such as the Bureau of Land Management have not been in the lead.[14]

The Nevada State Land Use Planning Council and the State Multiple Use Advisory Committee on Federal Lands agreed that Aerojet's environmental report was inadequate because it didn't consider mineral rights, access to public land, or water. They passed draft resolutions calling for a full environmental impact statement to be prepared on the project. They didn't feel that this would block Aerojet's expansion into Nevada. Rather, it would help local governments better understand the proposal.

The State Multiple Use Advisory Committee on Federal

Lands later rescinded its resolution. In response to charges that this was politically motivated, the committee chairperson replied that no political pressure had been applied. The resolution was rescinded because some members said that they might not have acted with full knowledge.[15]

The land swap attracted attention far beyond Nevada's borders. The 1987 World Wilderness Congress, which attracted 1,800 delegates from sixty nations, opposed the swap. The World Wilderness Congress focused on ways to balance economic development with preservation of natural resources. It passed a resolution urging that the Aerojet project pass all the tests of the National Environmental Policy Act and the Endangered Species Act. It challenged legislation in the U.S. Congress to bypass environmental protection measures in those laws, such as the requirement for an environmental impact statement.

The Nevada congressional delegation responded to public concerns by agreeing to hold public hearings. At a Senate Interior Appropriations Subcommittee hearing in Las Vegas, Nevada and Aerojet officials faced a "strong and well-coordinated opposition."[16] While most of the testimony dealt with the desert tortoise, other Nevadans focused on the planning process.

> The proposed land swap has a dark and unsavory history. It was worked up in the higher echelons of the Department of the Interior. The Bureau of Land Management was pushed aside and told to stay out of it. U.S. Fish and Wildlife Service biologists have been muzzled.[17]

Following the hearings, extensive negotiations between the congressional delegation and Aerojet resulted in a compromise bill that provided more protection for the desert tortoise. No environmental impact statement was prepared. Congressional staff correctly predicted the bill would pass due to support from Nevada officials.

On the day before April Fools Day in 1988, President Reagan signed the Aerojet land swap bill into law. Reagan said the exchange would expand the tax base and provide major employment opportunities in Mineral, Lincoln, and Clark counties. At the signing ceremony, Nevada Senator Chic Hecht said that the land swap could "be the beginning of thousands upon thousands of jobs in Nevada."[18]

The land swap act traded public land totaling 28,000 acres at Coyote Springs and 10,000 acres at Garfield Flat for 4,600 acres of Aerojet's property in Florida. It also included a lease of 14,000 acres of public land at Coyote Springs to Aerojet. Congress intended for the lease to protect desert tortoise habitat. Section 4 of the act stated that the lease had an "initial term of ninety-nine years, during which time no rental shall be required to be paid to the United States." It also gave Aerojet the right to sell the lease, subject to the approval of the secretary of the interior. The act required that Aerojet, in its use of the leased land, should follow the secretary's recommendations for protection of the desert tortoise "so far as possible."

The land swap act also prevented the possibility of any legal action under the National Environmental Policy Act. Section 10 stated that Congress believed that previous studies had been sufficient to meet the objectives of that act. Based on that finding it prevented any further legal action by any citizens "alleging a failure . . . to comply with any provision of law" other than the land swap act.[19]

Neither the fears of degradation of Nevada environmentalists nor the "chamber of commerce dreams" of Nevada officials have come true. Aerojet has abandoned its rocket facility plan and is trying to sell its Coyote Springs property. No destruction of desert tortoise habitat has occurred.

Aerojet built munitions bunkers on its Garfield Flat lands despite its promises that the lands would be used solely for buffer zones. Still, the land swap expanded the tax base, although not by as much as hoped. Lincoln County has netted about 50 cents an acre. Mineral County has only received 25 cents an acre.[20] The most significant impact has been the "No Trespassing" signs on what was formerly open public land.

The National Guard Tank Range

In December 1985, the Army Corps of Engineers filed an application for a right-of-way for 8,300 acres of land near Las Vegas to be used by the Nevada National Guard for tank training. The Guard

had been using the area under a land use permit issued by the
Bureau of Land Management. The permit was issued before the
Federal Land Policy and Management Act. That act eliminated the
authority to grant land use permits for military purposes.

The Bureau rejected the right-of-way application because it
was not a legal use. Congress intended for rights-of-way to be used
only for transportation or communication facilities. The Bureau
suggested that the corps consider withdrawing the lands under the
Engle Act.

The corps appealed the decision to the Department of the
Interior's Board of Land Appeals. It argued that years could be
spent waiting for a congressional decision on an Engle Act withdrawal and the National Guard needed the land immediately.

The board ruled in the BLM's favor, concluding that tank
training ranges were not a legal use for rights-of-way. It concluded
that the legal way to obtain a military range was through an Engle
Act withdrawal.[21]

This decision prompted the Nevada National Guard to search
for alternate sites. The Air Force rejected its request for joint use
of the Indian Springs Gunnery Range at Nellis. Attempts to use
the Army's National Training Center at Fort Irwin in southern
California were also rejected because the regular Army needed it
full time.[22]

In 1988 the Guard turned its attention to the Hawthorne area.
It developed an ambitious proposal to use Hawthorne Army Ammunition Plant facilities with 586,000 acres of adjoining BLM public lands and 24,000 acres of Navy-withdrawn lands for a national
Reserve Component Training Center.

The reserve component of the Army consists of the National
Guards of each state and the Army Reserve. Since the end of the
Vietnam War, the reserve component has increased rapidly while
regular forces have declined. The Army claims that technological
advances in weapons systems have greatly increased its need for
large training ranges. It had acquired most of its ranges during
World War II and the Korean War. The Army maintained that
those ranges were no longer adequate to train regular Army forces.

To meet its needs, the Army had applied for an Engle Act
withdrawal for 238,000 acres of BLM public lands to expand Fort

Irwin. It also planned to add 665,000 acres to Fort Sill in Oklahoma, 139,000 acres to Fort Bragg in North Carolina, 83,000 acres to Fort Polk in Louisiana, 82,000 acres to Fort Riley in Kansas, and 60,000 acres to the Yakima Firing Center in Washington.[23]

Despite these proposed expansions, the Nevada National Guard believed that the reserve component forces needed their own range. The reserve forces often have to schedule their training on a "space-available" basis. The Guard felt that this was not appropriate given its expanded role in national defense. It said that it needed a huge area because troops become too familiar with smaller ranges and training is less effective. The Guard claimed that not only was the Hawthorne site the "best location in the United States for desert operations,"[24] but the only land in the West where an entire armored division could train.[25]

The Montana National Guard disagreed. It found a better site. It proposed to obtain a thirty-year "land use agreement" with the BLM for 980,000 acres in northeastern Montana for a combined tank and bombing range. More ambitious than the Nevada proposal, it was intended to be used by all U.S. and allied military forces. The Montana Guard proposed to use the range for up to 15,000 troops and 1,500 tracked vehicles. There also would be air-to-air gunnery and air-to-ground bombing most of the year.[26]

The smaller Nevada proposal was intended to accommodate armored division-level training of 15,000 troops with 3,000 tracked vehicles. Guard units from throughout the nation would be invited. These troops would provide a major economic benefit to Hawthorne. Senator Reid said that the training center would provide millions of dollars annually for the local economy and 200 new full-time jobs. He obtained funding to initiate the project.[27]

Despite earlier failures, the Guard proposed to obtain the public lands for the range through a right-of-way grant. The Guard wanted to avoid a formal withdrawal because it would require congressional approval. Although Congress has often guaranteed non-military land uses through the withdrawal process, the Guard said that a right-of-way would allow miners, ranchers, and recreationists to continue using the land. It encouraged families to "come out and picnic and watch the guard go through its tank maneuvers."[28]

The proposal received the support of the Mineral County commissioners. One commissioner said that most of the county's residents supported the project because Hawthorne had been the site of defense activities since the twenties. The community had previously been supportive of military activities.

The reaction of other Nevadans varied considerably. The Sierra Club offered halfhearted opposition. The club's conservation chairperson suggested that a smaller range might be more appropriate and that the Guard should consider setting aside funds for wildlife habitat improvement. Others took stronger positions. The Nevada Outdoor Recreation Association contended that the Guard was trying to circumvent the Engle Act by applying for a right-of-way rather than withdrawal. It said it would "defend the Engle Act to the hilt," through lawsuits if necessary.[29] Others blamed Nevada's own politicians.

The ease with which Nevada's urban oriented elected officials allow the continuing allocation of rural land and air for military use, makes Nevada an attractive proposition for military projects which would be unacceptable in other states.[30]

The *Reno Gazette-Journal* charged that the Guard proposal was only another "military land-grab." It requested that the armed services make more than token interagency gestures for joint use of existing training ranges. It concluded that given the rapid population growth in Nevada, the state could no longer afford to accommodate the wishes of every branch of the armed services to establish its own training ranges.

In response to these concerns, the Guard decided to prepare an environmental analysis before taking any further action. It promised to hire an impartial consultant to do the environmental and economic impact studies.[31] The purposes of these studies were to decide if a full-fledged EIS was needed and if the proposal had any fatal flaws.

The Guard chose the Army Corps of Engineers as its impartial consultant to prepare the analysis. While this selection may appear questionable, the corps has established a reputation for solid, objective environmental analysis.[32]

The Army Corps of Engineers released the environmental assessment for the training range in September 1989. The assessment identified the Guard's site selection criteria:

Size: About 775,000 acres would be required for division-size maneuvers.

Shape: The potential site would need to be in a rectangular or oval shape to provide realistic training.

Terrain: The site should have a variety of mountainous, rolling and flat terrain.

Access: Since trainee units would be coming from all over the country, the site should be accessible by roadways, airfields and rail.

Training Mix: The area should have the potential to mix both ground maneuvers and aircraft operations.

Land Uses: Surrounding land uses should be compatible with tank training. No urban areas should be next to the site.

Environment: Since tank training is destructive, the site should have limited environmental and cultural resources.

Facilities: Existing physical facilities [such as the Hawthorne Ammunition Depot] should be near the site.

Five alternatives were analyzed in the document. The first was the National Guard's original proposal for 586,000 acres of BLM public land centered on the Gabbs Valley area northeast of Hawthorne. The second alternative was for a similar-sized area around the Pilot Mountains southeast of Hawthorne.

A third alternative involved joint use of the Nellis Range. The assessment noted several major obstacles, especially previous Air Force rejections. It also noted that since the Air Force had contaminated much of the Nellis Range by years of live bombing, it was unsafe for tank training. It also identified potential desert tortoise areas that could preclude tank training. (The federal government designated the tortoise an endangered species in the fall of 1989.)

The most significant alternative developed by the Corps of Engineers was the fourth. It proposed to conduct a nationwide study of all Department of Defense installations for training sites before any more military expansions would be considered. Nevadans had requested this measure for years without success.

Although the alternative was not fully developed in the environmental assessment, the Army Corps of Engineers recommended it be evaluated in detail if the National Guard decided to pursue its proposal. The corps thought that this alternative was the only way to respond to accusations that military ranges were automatically proposed for Nevada because of "its small population and the large proportion of federal public domain lands."[33]

The final alternative was no action. The Nevada National Guard and other reserve component units would have to continue training on a space-available basis at existing ranges. The corps concluded that this would result in less than ideal training for armored divisions.

The Army Corps of Engineers backed this discussion of alternatives with an objective analysis of environmental impacts. The assessment determined that significant adverse environmental impacts would result from establishment of the training range in either Gabbs Valley or the Pilot Mountains. Therefore, it concluded that further action on the proposal would require the preparation of a full-blown environmental impact statement.

The assessment projected that the most devastating impacts would occur to the region's fragile desert soils and vegetation. The assessment cited previous studies confirming the prolonged impact of tracked vehicles such as tanks on desert environments. Tracks from General Patton's tank training maneuvers during World War II are still visible in the southern California desert. Long-term desertification can result from extensive training maneuvers. The assessment noted that every previous study on the subject concluded that increased soil compaction occurred and that it was the "ultimate cause of far-reaching ecosystem damage."[34]

The assessment included an analysis of the destructive effect of tank training on desert vegetation communities. The assessment noted that the regeneration of shrub communities in high-desert environments takes several years. Regeneration of desert pinyon-juniper forests takes at least a century. It concluded that these communities were not recoverable on tank training ranges, noting that many experts doubt that heavily impacted deserts can ever be restored.

These direct impacts to the basic soil and vegetation resources were expected to lead to significant adverse impacts to wildlife due to habitat destruction. The assessment projected the degradation of recreation opportunities. The heavily disturbed soils would be vulnerable to wind erosion, causing dust storms. On the other hand, economic benefits to Mineral County from increased military spending were foreseen.

The assessment reported that cumulative impacts were also an issue. The training range proposal along with other military proposals for increased land and air space use would result in an expanded military presence throughout Nevada.

The assessment identified five potential Navy land and air space acquisitions in the region. They included the 181,000-acre Fallon Naval Air Station Master Land Withdrawal and a conceptual plan for an additional 240,000 acres for a new Navy bombing range. The Navy had also applied for a new military training route over Walker Lake and Hawthorne for up to 100 flights per day. The Navy had proposed another route for testing Tomahawk cruise missiles. The missiles would be launched off the Pacific Coast, fly east of Hawthorne and land on the Navy's Fallon bombing ranges. Long-range air space proposals called for the creation of massive new military operating areas starting east of Hawthorne and extending nearly to the Utah border.

The corps included a frank discussion of legal and political issues to be considered by the National Guard. It warned that the BLM had rejected other proposals for rights-of-way for military maneuvers. It also cautioned against attempting to get a lease or permit for the training range from the BLM. Federal agencies are prohibited by law from holding leases or permits. Although it could be argued that the National Guard is a state agency, the corps concluded that tank training was a *federal* national defense activity. Due to the high sensitivity of Nevadans about increased military use of public lands, the Army Corps of Engineers advised against any questionable means to obtain the range. It alerted the National Guard to the potential for litigation by opponents of greater military land use.

The corps recommended that if there were support from Nevada's congressional delegation, an Engle Act withdrawal should be pursued. The only legally defensible alternative would be to seek special legislation outside the Engle Act process. The corps made only one major error in the environmental assessment. It said that such legislation would be subject to the requirements of the National Environmental Policy Act. Actually, special legislation is subject only to constitutional requirements and can avoid compliance with environmental laws.

Since the environmental assessment concluded that much more detailed analysis would be required in an environmental impact statement, the Nevada National Guard decided not to hold a public hearing on the assessment. It did ask for public and other agency review of the document.

The Bureau of Land Management agreed with the general conclusions in the assessment. But it outlined the Montana National Guard proposal for a similar training range and suggested that it wasn't practical for every state to have its own massive armored division training center. It concurred with the corps' suggestion for a nationwide study of training needs and suggested that it consider minimizing the taking of public land. Most important, it recommended that the study provide "citizen involvement as early in the planning process as possible."[35]

Citizen Alert, a Nevada-based military watchdog group, wanted a more comprehensive planning effort.

> Although the Environmental Assessment for the Reserve Component Training Center states that the need for the facility derives from a national level problem, the proposal was developed by the staff of the Nevada National Guard. The question was simple—"Where in Nevada can the RCTC be placed?"—rather than outlining the actual problem and looking at possible solutions.[36]

Citizen Alert called for a complete and quantitative survey for threatened or endangered species. It identified the areas proposed for the training range as rich in mineral deposits, advising that conflicts between the National Guard and miners were inevitable. It recommended that the Guard conduct an inventory of the region's mineral resources before proceeding with its proposal. Finally, it warned that any attempt to circumvent the Engle Act by trying to get the land through special legislation would be opposed.[37]

The Nevada and Montana National Guard proposals and other Army, Navy, and Air Force plans attracted the attention of the national media. The *Wall Street Journal* charged that the proposals were "piecemeal, uncoordinated and redundant."[38]

In 1988, the House Interior Committee took up the issue of National Guard proposals. Neither the Engle Act nor FLPMA discussed procedures for permitting state military departments to use federal lands. The Bureau of Land Management and the Forest Service had been permitting various state-level military activities under a variety of authorities with no consistent approach. In some cases the bureau had rejected state military uses, requiring Engle Act withdrawals through federal military agencies such as the Army Corps of Engineers. Public concern had developed because these decisions were made within the executive branch and lacked congressional oversight.

The House Interior Committee approved of a bill to have state military agencies follow the same procedures as federal military agencies. The most important part of this bill was the requirement for congressional approval of withdrawals over 5,000 acres. In effect, the bill was an Engle Act for state military departments.

The Nevada National Guard unintentionally made a strong case for this type of bill in its justification for the Hawthorne training center proposal. The Guard said that the Department of Defense's new "total force" concept meant a greatly expanded role for the National Guard. It more fully integrated the Guard with the regular Army. The Guard was also modernizing its equipment to a level nearly identical with the Army's. Therefore the Guard needed to train more like the regular Army.[39]

The House Interior Committee came to similar conclusions. It decided that the distinction between state and federal military withdrawals was "not a sound one for purposes of land management" and that more consistent standards were needed.[40]

Nevada Representative Vucanovich and others disagreed. They argued that congressional approval of National Guard withdrawals would result in increased costs and delays for the Guard. They saw a major difference in training activities between the

Guard and the regular Army and didn't think that a uniform standard with federal withdrawals should be required.[41]

The opponents succeeded when the full House acted on the bill. The proposal for applying Engle Act procedures to state military actions was watered down by several amendments proposed by the National Guard. These included authorizations for National Guard proposals that would not result in significant residual contamination of public land. There were no limits on the number of acres that could be approved, and these authorizations would be limited to three years, but could be renewed indefinitely. Even in cases where significant residual contamination would occur, approvals for up to 10,000 acres could be granted. Only withdrawals with significant contamination of more than 10,000 acres would follow Engle Act procedures. The amendments also provided that federal military departments—the Army, Navy and Air Force—could use the state-authorized lands without congressional approval. The bill is currently under consideration by the Senate.[42]

Meanwhile, Kansas Senator Nancy Kassebaum requested a General Accounting Office study of Army land acquisition proposals. While she was most interested in a proposal to expand Fort Riley in Kansas, the study evaluated overall Army procedures. It concluded that rather than proposing land acquisitions through a rational planning process, the Army focused on "targets of opportunity." One of those targets specifically identified in the study was the Fort Irwin withdrawal. Most acquisitions have been the result of actions by commanders of individual installations rather than through an Army-wide analysis of priorities.

Army planning procedures also limited real up-front consideration of alternatives. According to the study, the Army makes initial decisions on land acquisitions before it analyzes any alternatives. This limits its ability to seriously consider increased use of existing withdrawals or greater reliance on simulators.[43]

In response to concerns by Chairman Vento, the House Interior Committee reopened the entire subject of military land withdrawals with hearings in January 1990. The committee noted that demands for public land from both federal and state military agencies amounted to more than 4 million acres in Nevada, Utah, Idaho, Montana, California, Colorado, and Washington. Although the hearings have not yet resulted in any firm legislation, the committee was able to hear testimony from a variety of federal agencies and citizen representatives. This included the testimony of Citizen Alert.

Circumstances have never been more favorable for the initiation of solutions to the military abuse of public lands, people and wildlife. The dramatic changes in superpower relationships, the growth and success of democratic movements in eastern bloc countries, and the general easing of tensions over the threat of nuclear war should help satiate the Pentagon's appetite for more land and airspace. Indeed, these events do not lend credibility to the Pentagon's plans for its huge buildup in the western United States.[44]

The Nevada National Guard put its training range proposal on indefinite hold in April 1990. While the environmental assessment and subsequent public comments likely had an effect, the decision was the first major impact of the end of the Cold War on Nevada. The Guard's Nevada commander, General Clark, cited social and political changes in Eastern Europe that could lead to major reductions in Army forces. He said if reductions occur, reserve component troops could use Army training ranges in California and Arkansas. He concluded that the military "shouldn't be running off willy-nilly trying to get a new training center until we see if it's needed."[45]

SEVEN

Military Planning Doctrine

The impacts of perestroika, glasnost, and the profound changes in Eastern Europe on American military doctrine should not be underestimated. For forty years American military policy has been based on a perceived Soviet threat. Political and military doctrine can no longer be so clearly defined. The idea of Russian tank battalions pouring through the Fulda Gap into Western Europe is becoming as antiquated as the domino theory in Central America. The Middle East is likely to be the primary focus for future military doctrine.

Yet the armed forces in general, and the Navy in particular, are slow to react to outside influences. The phrase "the Navy Way" did not come about by pure chance.[1] Franklin Roosevelt once said that his Treasury Department was so ingrained that it was impossible to make any changes. Even more rigid was the State Department with its career diplomats. But even it was far from his toughest adversary:

> But the Treasury and the State Department put together are nothing compared with the Navy. The admirals are really something to cope with—and I should know. To change anything in the Navy is like punching a feather bed. You punch it with your right and you punch it with your left until you are finally exhausted, and then you find the damn bed just as it was before you started punching.[2]

Recent military land use proposals in Nevada were developed under the policies of the eighties. Though those policies may soon be outmoded, it is important to understand the philosophy behind them.

The word "planning," as used by the Department of Defense, means the establishment of goals, policies, and procedures to provide for the defense of the nation. As such, it focuses on issues like weapons development and force structure rather than the traditional urban planning focus on the built environment. Land use planning is a minor part of overall military planning, but has had a major impact in Nevada.

The framers of the Constitution feared the power of a standing army, yet were concerned about the effectiveness of Congress to manage the military. Therefore, they placed the military in the executive branch with a civilian president as commander-in-chief. The founding fathers provided a limit on the power of the military by leaving funding authority with Congress. This system remained in effect for only nine years, when Congress split the military into a Department of War and a Department of the Navy. That basic structure held until after World War II.

Current military planning doctrine and structure was shaped by congressional and executive branch decisions in 1947 and 1948. Congress created a Department of Defense in 1947 to oversee three services—the Army, Navy, and a newly created Air Force. In 1948, the secretary of defense and the Joint Chiefs of Staff signed the Key West Agreement, which staked out the turf for each service. The agreement included, for example, a provision that prohibits the Army from developing a fixed-wing aircraft capability. As a result, the Army developed helicopters that emulate fixed-wing capabilities. The agreement set the conditions under which Army, Navy, and Air Force planners compete for programs, funding, and power.[3]

Interservice competition is supposed to take place under a framework established by the Department of Defense to deal with the problems of nuclear war strategy and development of new weapons systems. But other issues have often dominated the debate.

The controversies that most intensely engaged their pro-
tagonists were the quasi-theological questions of roles and
missions: Were missiles a new form of artillery, and therefore
within the proper domain of the army, or a new kind of air-
plane and, therefore, the preserve of the air force? Could the
army properly use close-support aircraft painted army colors
and flown by pilots in army uniforms, or should their planes
be painted air force colors and flown by pilots in air force
uniforms?[4]

There is no doubt that the armed services use such arguments
to further their own interests. Yet, each service truly believes that
its interests coincide with the national interest.

Competition between agencies is common at all levels and in
all functions of government. It is not necessarily negative. For
example, there is a natural rivalry between the U.S. Forest Service
and the Park Service since most national parks are carved out of
the forests. Competition has improved both agencies as they strive
to demonstrate their expertise. This was particularly apparent in
the development of imaginative competing recreation strategies
during the 1950s.[5]

Similarly, competition between the armed services has many
benefits. Competitive innovation leads to rapid development of
new defense strategies and weapons systems. Still, the attention
given to problems with interservice rivalry is probably warranted.
It leads to many wasteful excesses. An example is military intelligence. The Defense Intelligence Agency (DIA) was created to con-
solidate the views of the various service staffs and eventually
replace them. When the intelligence staffs were transferred from
the military departments, they built up staffs in areas such as
"foreign technology." This allowed them to compete with DIA as
well as each other. The Air Force intelligence service alone em-
ploys about ten times as much staff as the DIA and outspends it by
25 to 1.[6]

Interservice competition is complicated by the lack of a level
playing field. A disparity in the bureaucratic resources of the services can have more influence on defense decision making than the
actual merits of the competing decisions. The Navy has a distinct

advantage in this arena. The Navy, more than the other services, is an institution. It has a strong sense of tradition and a clarity of purpose. It is the only service with its own navy, army, and air force. This makes it the least dependent on the other services. The Navy has a more elaborate, hierarchical structure than the other services. Carrier-based fighter aviation is at the top, followed by attack or bomber aviation.[7] Land use planning is much lower in the hierarchy.

The Navy's strengths as a bureaucracy were reflected in budget increases during the military expansion of the 1980s. It was also reflected in Nevada. The Navy pursued land and air space withdrawals more aggressively than the other services.

In such an environment, efficient resource allocation is difficult to achieve. Overuse of the nation's land resources by duplication of training ranges can result. Nevadans have contended for years that the military should use the Nellis Range more efficiently before seeking new withdrawals in their state.[8]

It has often been argued that more centralized control of the services is needed, yet Congress purposefully limited the size and power of the Department of Defense and the joint chiefs' staff in relation to the services. These limits were intended to perpetuate interservice competition because Congress feared the power of a strongly centralized Department of Defense. The overall size of the military establishment that a superpower has needed to preserve an international power equilibrium could be beyond the capacity of a democratic government to control. Therefore, competition between services may help to maintain civilian control.[9]

Other than general missions, the most intense competition has been focused on new weapons systems. Military planning doctrine has been based on the view that military superiority should be attained through technological superiority. This has led to an increasingly complex and expensive force structure.[10] As the complexity of weapons systems has increased, the need for training has increased. This has led to increasing demands for training ranges.

These demands accelerated rapidly in the eighties as defense spending mushroomed. The perception that the Soviet threat was so great that any amount of spending was justified was common. In such an atmosphere, the identification of priorities was meaningless, including priorities for land uses. The administration used spending to show commitment as much as to generate specific results. This was done in the framework of a broad planning doctrine: that is, U.S. military forces should be able to defeat any enemy force anywhere on the planet. This had direct implications for the development and acquisition of new weapons systems.

Since this strategy is broadly framed and without specific guidance (i.e., "be prepared to fight anything, anywhere"), the result during the first four years of the Reagan buildup was predictable and observable: As many weapons as possible, of every type available, were acquired on the theory that some situation would be encountered where each weapon would be the key to victory.[11]

Actual long-range military planning is directed by Congress. Decisions by Congress to fund new weapons systems or new numbers of existing systems have direct impacts on the armed services' demands for land and facilities. Each of the services has developed land use planning systems to respond to changes in the number and type of weapons systems.

Although the differences among the armed services should never be minimized, a look at Navy planning procedures can reveal general military planning policies. Lieutenant Keefer, a character in Herman Wouk's *The Caine Mutiny*, described the Navy as a master plan designed by geniuses for execution by idiots. He meant that everything the Navy does is outlined in meticulous detail. Individual tasks are clear to those responsible for their execution. Navy master planning is no exception.

The Naval Facilities Engineering Command (NAVFAC) is responsible for the planning, construction, and maintenance of naval shore facilities. Until the development of naval air capabilities, these functions were limited to bases for the U.S. Fleet. Now its role includes continental bombing and test ranges. NAVFAC's planning function involves conducting studies and analyses of existing facilities, new requirements, and the implications of civilian developments to Navy activities.

NAVFAC is headquartered in Washington, D.C., under the Chief of Naval Operations. Its western regional headquarters is located in San Bruno, California. It is responsible for land use planning for all states on the West Coast as well as Nevada, Arizona, Idaho, and Utah. During the mid-1980s its total employment reached nearly 1,000. The majority are professional civilian engineers, architects, real estate specialists, and administrative staff. About twenty-five are full-time professional planners. Each is assigned to one or more of the nearly 100 Navy installations in the western division.[12]

NAVFAC carries out its planning mission on several levels. The broadest is regional and systems planning. This involves analyzing Navy activities such as pilot training that are related by common geography or as elements of an integrated system. The Navy examines these interrelationships to help determine facility needs such as additional bombing ranges.

Master planning is focused on an individual base or facility. In many ways it is similar to local government master planning. It is clearly based on the rational-comprehensive model of planning. The Navy defines master planning as the scientific art of comprehensive planning for the development of its facilities. It is supposed to include consideration of the total environment: physical characteristics, operational necessities, human interests, natural resources, and areas of mutual interest beyond station boundaries.[13]

A master plan's major purposes are to determine the best use of existing land and to identify the need for additional land. A NAVFAC master plan includes descriptions of existing conditions, development and operations constraints (such as adjacent community development), and Navy mission requirements. It proposes future development goals and includes capital improvement plans for projects needed to implement the master plan.

Master plans are intended to overcome the problems associated with frequent turnover of individual base leadership. The Navy's program for rotation of command can lead to major continuity problems. Master plans are designed to limit frequent policy changes based on the individual preferences of base commanders.

Beginning in 1973, the Navy began its Air Installation Compatibility Use Zone (AICUZ) planning process. It was intended to control encroachment of civilian activities in adjacent communities on Navy air operations. This was the result of the closure or scaling down of several Navy air bases because of off-station residential or commercial development. Like civilian airport planning, it focuses on preventing incompatible uses such as residential development near enough to Navy air bases to generate aircraft noise and safety problems. The Navy has developed AICUZ studies for all air bases and Range AICUZ studies for all bombing and test ranges. The AICUZ process includes developing agreements with surrounding communities to insure compatibility among land use plans. This includes Navy proposals for local zoning that prohibits residential development next to airfields.

The Navy has been concerned not only about residential development in populated areas, but also about other types of development in rural areas. Its concerns include highway construction or recreational facilities that could act as magnets for other types of allied developments and increased population near Navy air operations.

The Navy feels it should have the right to exercise control of off-base development because in many cases where there has been an encroachment problem between a military base and a civilian community, the military base was there first. Community contacts are an important part of AICUZ plans, but unlike the trend in local government planning, these contacts are generally one way. That is, they involve the Navy telling the community what its plans are rather than having formal public involvement in the preparation of plans. The Navy stresses the need to make civilian planning authorities fully aware of its own mission requirements, but limits opportunities for formal public involvement in the development of plans.[14]

Although the Navy focuses AICUZ plans on noise and safety issues, it considers other factors, such as the impact of off-site development on the quality of Navy training activities. This can lead to Navy requests to prevent the building of livestock fences near its bombing ranges because pilots could use them to help focus on targets, decreasing the effectiveness of their training. The outcome of the AICUZ process may be the identification of a need to acquire more land, as in the Fallon Master Land Withdrawal proposal.

NAVFAC also produces special planning studies for short-term site-specific projects. Often these are studies that individual bases would ordinarily prepare, but are referred to NAVFAC due to the level of controversy or complexity. Recently completed special planning studies include:

A Naval Materials Command colocation study in San Diego to decide which organizations would benefit from being located in the same area.

Waterfront planning at the naval complex in Seattle to support fleet expansion.

An examination of air support facilities at Las Maddalena, Italy, to develop a recommendation for military construction and special projects.

The Naval Facilities Engineering Command is also responsible for developing natural resource plans for the various installations. It is Navy policy to provide for the conservation and use of the natural resources under its jurisdiction. This includes a mandate for the development of plans for fish and wildlife habitat management that are compatible with state fish and game regulations. Public access to Navy property is supposed to be provided except where incompatible with security or public safety.

NAVFAC develops all of its planning efforts under the general guidance of the Navy's Shore Facilities Planning System. The system is intended to identify facilities needed to directly support a Navy mission. Facility needs are supposed to be consistent among similar Navy activities. All existing Navy properties are inventoried. Mission requirements are compared with existing facilities to decide if additional needs can be met with excess property. NAVFAC then identifies valid projects and submits budget requests. Excess property is supposed to be transferred to other agencies or Navy activities. The planning system provides for the acquisition of real estate based on military requirements and the need to protect investments in existing property.

At the local base level, land use planning activities fall within the purview of the public works office. Other functions include real estate management and facilities repair. Among these activities, planning has not been given much precedence; it is often an "underfunded, low priority function." At the operational base level, the Navy views it as a nonproductive element.

One tenet of organizational thinking in Public Works is to keep "planners" away from "doers" to avoid conflicts of interest and to free the productive element to devote its entire effort to "doing."[15]

The facility planning officer of the public works staff generally carries out planning at the base level. Planning officers are generally ensigns with no previous experience. The position is intended to help junior engineering officers become familiar with the Shore Facilities Planning System.

Part of the planning function is liaison with civilian planning agencies. The Navy has recognized that it is part of a larger community and is increasingly subject to local and state statutes regarding pollution control, building codes, planning ordinances, and traffic regulations. It also recognizes that there is a tendency within the Navy to view each base as an island. The Navy views this as an aberration that must be overcome by developing an attitude that it functions within a broader community.

The Navy made a real effort during the 1980s to improve community involvement in its planning efforts. In the 1970s, it provided public involvement only when required, i.e., for environmental impact statements. Navy planners now provide for informal public involvement on an ad-hoc basis. They recognize that it improves chances to obtain military objectives. Yet public involvement is still not recognized as a fundamental right, and the attitude that the Navy knows best remains. Citizen involvement in Navy decision making is only beginning to be accepted.

Navy careers are based on a highly structured promotion system. Tolerance for error is minimal, and there is a very strict adherence to deadlines. The Navy is a very product-oriented organization, and missing one deadline can overshadow years of quality work. Since it can lead the planning process in unexpected directions, public involvement is viewed with fear and is avoided if at all possible.[16]

EIGHT

Case Study
Expanding the Bravo Ranges

This case study is an analysis of the Navy's Fallon Master Land Withdrawal to expand its bombing ranges. It focuses on the process used to develop the withdrawal proposal and on public involvement in that process.

In late 1978, the BLM notified the Fallon Naval Air Station (NAS) that geothermal developers were taking an active interest in public lands near the Bravo 19 range (Map 5). Fallon NAS staff believed that this could be a major threat to their pilot training operations. Geothermal production involves the use of drill rigs more than fifty feet high. Producing wells are flow tested, releasing large steam plumes. Many geothermal power plants also release significant amounts of steam during general operation. The Navy thought that this would interfere with its operations on Bravo 19 by creating safety hazards for its pilots.

Fallon NAS staff developed a proposal for a buffer zone on the approach to Bravo 19 to control any potential geothermal operations. During this process, they identified another major problem at Bravo 19: pilots were occasionally missing targets, delivering or skipping ordnance off-range. So they decided to expand the proposed buffer zone down-range from Bravo 19. This led them to take a close look at the other ranges at Fallon, and they developed proposals for buffers to prevent geothermal development and public access to areas where bombs were dropped off the Bravo 16 and Bravo 17 ranges as well.[1]

Naval air station staff also decided to apply for more control

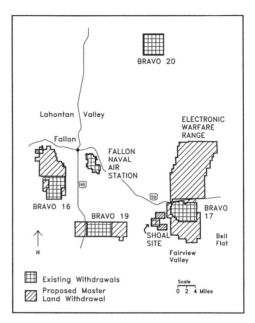

Map 5. Fallon Naval Air Station Land
Withdrawals

over the electronic warfare (EW) range and the Shoal Sites near
Bravo 17 (Map 5). The Navy used the 37,000-acre EW range to
provide electronic threat emissions for aircraft using Bravo 17. The
Shoal Sites consisted of three adjacent parcels totaling 7,400 acres,
used by the Navy for helicopter search-and-rescue training. The
central 2,500-acre site was withdrawn by the Department of Energy. It conducted underground nuclear testing there during the
fifties. The other two parcels were used under a special land use
permit from the BLM.

The Federal Land Policy and Management Act limited use
of BLM-managed lands by other federal agencies to withdrawals,
rights-of-way, and cooperative agreements. It limited rights-of-
way to water systems, pipelines, utility lines, communication sites,
roads, railroads, and other transportation systems. The threat
emitter sites on the EW range were deemed communication sites
and subsequently granted rights-of-way.

FLPMA only allowed cooperative agreements "where the
proposed use and development are similar or closely related to the

programs of the Secretary [of the interior] for the public lands involved." The BLM decided that the Navy's uses of the EW range and Shoals Sites were not related to any Department of the Interior programs. Therefore, it canceled permits for the Navy's EW range and the two BLM Shoal Sites. The Navy retained control of the central Shoal Site, since it was authorized under Department of Energy regulations. Fallon NAS staff decided to try to reestablish Navy control over the lands it had lost along with the buffer zones identified for the other ranges.

The Navy also gave major consideration to the proposed basing of F/A–18 aircraft at Fallon. The most commonly used attack aircraft in the early eighties were the A–6 Intruder and the A–7 Corsair. Both planes were designed for subsonic speeds. The F/A–18 Hornet has replaced most of the A–7s. It is a dual-mission aircraft that can function as a fighter or a bomber. Its other primary feature is speed: The F/A–18 is supersonic with a top speed of more than Mach 1.8. When loaded with bombs it flies at subsonic speeds because of the increased drag of the ordnance. Once the F/A–18 has dropped its bombs it can go supersonic and defend itself as a fighter, since its standard loadout includes air-to-air missiles. It can also be equipped totally as a fighter, with air-to-air capabilities only.

The move to F/A–18s called for a major increase in pilot training. The Navy proposed a $350-million simulator program using the most advanced technology available, but nevertheless projected a major increase in hands-on training at the Fallon Naval Air Station.[2] Not only were more training flights projected, but the flights would be at greater speeds. Slight pilot errors at these speeds could result in missing targets by a wide margin. Therefore, expanded buffer zones around the targets were believed necessary.

Fallon staff refined the master land withdrawal proposal and sent it up the Navy's chain of command in May 1979. It included five areas. A synopsis and brief background of each is given here.[3]

Bravo 16. The proposed buffer zone includes 34,000 acres north of the existing 17,000-acre range. Most of the lands were previously withdrawn by the Bureau of Reclamation as part of the Newlands reclamation project. The area generally consists of saltbush flats and sand dunes. The Truckee Carson Irrigation District manages most of it for the Bureau of Reclamation. It contains canals and other water distribution facilities as well as Sheckler Reservoir, a main water storage facility for the Newlands project. The reservoir provides habitat for many waterfowl species and is a popular recreation area for Fallon residents. The Navy's primary concern was to prevent any development, including geothermal, from interfering with the low-level run-in to the Bravo 16 target area. Residential growth from Fallon has been moving in the direction of Bravo 16, and many noise complaints about use of the range have been received.

Bravo 17. The proposed buffer zone totals 28,000 acres to the south of the existing 21,000-acre Bravo 17 range. The lands consist primarily of shadscale and greasewood flats in southern Fairview Valley. Some limited mining activity has also occurred. The primary use of the lands is for livestock grazing. A paved road, State Route 31, is located in the buffer zone and is used by ranchers, miners, hunters, and other recreationists. The Navy's purpose for the buffer zone was primarily for safety reasons. Live ordnance had been delivered or "splashed" to the south of the existing range.

Bravo 19. Two proposed buffer zones, located west and east of the existing 17,000-acre range, include 20,000 acres. The west buffer zone includes part of U.S. 95, the major highway from Fallon to Hawthorne and south to Las Vegas. This parcel is located on shadscale and greasewood flats on the eastern slope of the Desert Mountains. It is occasionally used for hunting and recreation access to the Desert Mountains, although its primary use is for livestock grazing. This was the first parcel identified as a buffer zone to prevent geothermal development from interfering with range operations. The eastern buffer zone includes the Blow Sand Mountains and upland sagebrush areas of the Cocoon Mountains and Barnett Hills. It contains part of Diamond Jack Wash, an area used for livestock grazing, and includes a major livestock watering facility. Mining has occurred in the Blow Sand Mountains. As live ordnance has been delivered or skipped to the east of the existing range, the primary purpose of this part of the buffer zone was public safety.

EW Range. This area consists of 37,000 acres previously controlled by the Navy under a special land use permit that was subsequently canceled due to provisions in the FLPMA. It also includes a buffer zone totaling 57,000 acres north of the old EW range. The area consists primarily of shadscale and greasewood flats with significant stands of Indian ricegrass. It also includes the southeastern slopes of the Stillwater Mountains, including a portion of the BLM's Job Peak Wilderness Study Area, and part of the Louderback Mountains, including the Wonder Mining District. A portion of a paved county road that provides the primary access to Dixie Valley is also included. The primary uses of the area are livestock grazing and Navy electronic warfare sites. It contains roads used for hunting and recreation access to the southern Stillwater Mountains and the La Plata areas. The Navy wanted to regain control of the EW range and provide a buffer zone to prevent geothermal and mining development in the low-level run-in corridor north of the EW range.

Shoal Sites. This proposal includes two parcels totaling 4,600 acres located on either side of the existing Department of Energy site in the Sand Springs Range. They are located in an area that has been the subject of a major joint Bureau of Land Management/Nevada Department of Wildlife project to develop chukar partridge habitat. The area is heavily used for chukar hunting. Livestock also graze here. The Navy proposed to regain control of these sites after its land use permit was canceled following the passage of FLPMA. The Navy wanted to have the authority to prevent mining and other forms of development that could interfere with its search-and-rescue training.

After the Navy Department approved the request to proceed with the withdrawal proposal, Fallon NAS staff prepared a requirements analysis. Briefings were held for the Churchill County Commission. The major public contacts were with the Bureau of Land Management, and they focused on whether the Navy could control activities on public lands without a formal withdrawal.

The Navy pursued other administrative actions because it was aware of the potential controversy associated with formal withdrawal. In addition, it did not feel that it had enough time to go through the withdrawal process given the immediate potential for geothermal development in the area. A geothermal resource exists about a mile north of Bravo 19 at Lee Hot Springs. Little is known about it. The resource is classified as having speculative potential only. A more serious prospect is located in southern Dixie Valley near the electronic warfare range. Three deep exploration wells were drilled in the early 1980s, apparently without success. No further exploration has occurred. Oxbow Geothermal Corp. has developed the largest geothermal plant in Nevada in northern Dixie Valley, thirty miles to the north, but has no plans for the southern Dixie Valley area.

The Navy's preferred options to avoid geothermal development were either rights-of-way or cooperative agreements to restrict land uses. However, neither of these options would control mining development on public lands. The 1872 Mining Law is still in effect. It allows miners to obtain legal patent, and therefore control of BLM-managed public lands, if they can prove a valid mineral deposit. Even on lands remaining in public ownership, any restriction of mining activity beyond preventing unnecessary degradation would constitute a taking of private property rights. The Navy eventually reached the conclusion that its only option was formal withdrawal.

The RAICUZ Study

In 1980, the Navy initiated a Range Air Installation Compatibility Use Zone (RAICUZ) study. This was a detailed planning analysis prepared by the Naval Facilities Engineering Command's western headquarters in San Bruno, California, under Department of Defense regulations. Completed in 1982, the RAICUZ study formed the technical basis for all subsequent Navy actions regarding the master land withdrawal.

The Navy contacted Churchill County, the BLM, and the Bureau of Reclamation staff and provided formal review for state planning staff. Yet, the RAICUZ process did not include any formal opportunities for citizen input or review. It was an internal Navy planning process prepared by experts in Navy procedures for examining land uses as they relate to the Navy's need for training ranges.

The study focused on three primary issues: (1) an increase in residential land use in the Lahontan Valley near Bravo 16 and in Dixie Valley near the EW range, (2) exploration and potential development of geothermal resources, and (3) potential designation of BLM wilderness areas near the EW range.

The study outlined the reasons why the Bravo ranges were particularly suitable for bomber pilot training.

The terrain itself is a singular combination of flatlands, mountains and canyonlands, which provides excellent ground conditions for practicing various types of aerial maneuvers, from bombing and strafing approaches to detection avoidance. The dry climate and clear air provide excellent visibility for most of the year. Population is very low, and entirely unpopulated areas exist which allow for safe practice maneuvers of drops, including use of live ordnance at the most remote ranges.[4]

The RAICUZ study focused on the complex land management situation at the Bravo 16 range. It noted that the Churchill County Commission controlled land use planning and zoning for adjacent private lands. The state planning coordinator served as a clearinghouse for planning federally managed lands. The Truckee Carson Irrigation District managed adjacent lands withdrawn by the Bureau of Reclamation from the BLM. Other public lands were managed directly by the Bureau of Land Management. Land management around the other ranges was considerably simpler, with most under the jurisdiction of the BLM.

RAICUZ studies involve computer modeling of noise and safety zones. The Navy mapped three noise zones for each range. They were based on decibel levels of 65 and below (Zone 1), 65 to 75 (Zone 2), and 76 and above (Zone 3). Safety Zone A, extreme hazard, was centered on the major bomb targets; Zone B was the area of armed overflight; and Zone C consisted of minimum hazard areas under restricted military air space. Navy planners overlaid these to identify incompatible land uses for Bravo 19 and the other ranges (Map 6).

The RAICUZ study generally confirmed conclusions about each range outlined in the Navy's original withdrawal proposal.

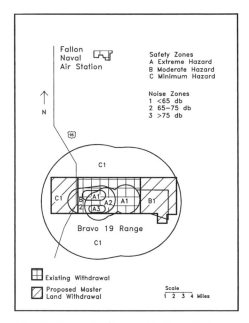

Map 6. Bravo 19 Range Air Installation
Compatibility Use Zone Study

Those conclusions were that noise and safety buffer zones were
needed around existing bombing ranges and additional withdrawals
were needed for electronic warfare and search-and-rescue training.
Several minor changes were made, including the identification of
additional safety hazard areas around Bravo 16 and Bravo 17.
These resulted in the modification of the proposed withdrawal
boundaries to include areas on the southwest and southeast boundaries of Bravo 16 and the west and east boundaries of Bravo 17.

Short-term (one-year) RAICUZ recommendations involved
Churchill County planning and zoning and BLM land management.
The Navy asked the county to pass a "truth-in-sales" ordinance
to make sellers in Lahontan and Dixie valleys disclose that property was subject to military overflights. It requested a zoning
ordinance restricting residential development to a minimum of
five-acre lots in areas near Bravo 16.

Restrictions on geothermal exploration and development were
proposed for BLM lands. The Navy wanted a new BLM public

land use category for buffer zones beneath military operating areas and opposed wilderness designations for BLM lands.

Midterm (five-year) recommendations were essentially contingency plans if the short-term measures were ineffective. They involved changing run-ins and mission types for Bravo 16 and the EW range in order to reduce overflights of residential areas. In the long term, more drastic measures were considered.

As a last resort, in the unlikely case that all ongoing, short and mid-term actions cannot result in a generally compatible future environment, NAS Fallon and higher commands should look at moving one or more of the ranges. Such a move would require extensive analysis regarding the feasibility and justifications for the move, and the impacts on the city and county.[5]

The Master Plan

The Naval Facilities Engineering Command's Western Division prepared a master plan for the Fallon Naval Air Station at the same time it was working on the RAICUZ study. The planning process consisted of six steps:

1. Scoping.   A questionnaire was completed by all Fallon NAS departments and divisions, returned to NAVFAC in San Bruno, California, and analyzed by the planning staff.

2. Inventory.   Land use data were compiled and analyzed. They were based primarily on two weeks of field investigations at Fallon Naval Air Station.

3. Alternatives.   The planning data were analyzed by NAVFAC planners to develop a number of alternative development concepts. These alternatives were then narrowed by Fallon NAS staff.

4. Draft.   Specific site designs and facilities were outlined in a draft master plan.

5. Review.   Fallon and Pentagon commands completed a detailed review.

6. Final.   Based on that review, NAVFAC completed the final master plan.

During this process, no formal public hearings were provided. The master plan focused primarily on physical improvements to the air station complex. It called for runway improvements, development of an integrated shopping and entertainment center, improvements to ordnance-handling areas, a new traffic circulation plan, and new enlisted personnel housing. Many of these improvements were designed to accommodate the proposed detachment of F/A–18 aircraft to Fallon. The master plan included a detailed capital improvements plan for the Bravo ranges, consisting primarily of new electronic warfare equipment. It also included the master land withdrawal.

The Withdrawal Petition

The withdrawal process is governed by public land regulations. They require the submission of a formal withdrawal petition. It must contain an analysis of why other administrative mechanisms cannot be used and why alternative sites are inappropriate.

The Navy submitted its petition for the master land withdrawal in September 1982. The BLM then issued a *Federal Register* notice describing the purpose of the withdrawal as a low-level, high-speed weapons delivery maneuvering area in connection with live ordnance drop.

The notice segregated the lands from operation of the public land laws for a period of two years. In effect, a segregation is a temporary withdrawal while the formal withdrawal is being analyzed. Jurisdiction remains with the Bureau of Land Management. The primary purpose of the segregation is to prevent nuisance mining claims from being filed in the area. The notice also outlined the bureau's role in the withdrawal process. It stated that the BLM would negotiate with the Navy to ensure that the proposal was for the minimum area necessary to meet the Navy's training needs. The proposal was intended to provide for the maximum public use of the withdrawn lands through an agreement on joint BLM-Navy management.

Four years after the Navy initiated the withdrawal proposal, formal public involvement was provided. The notice called for public comments in writing and announced a public hearing. This announcement came on the heels of the MX missile controversy when antimilitary sentiment was high. The Navy chose an inopportune time to publicize such a controversial project. This announcement was also accompanied by another Fallon NAS proposal to establish a supersonic operations area over central Nevada to accommodate the new F/A–18s. The public controversy they generated was no surprise. Comments opposing the withdrawal came from the same general coalition opposed to the MX, including the Nevada Cattlemen's Association, Nevada Woolgrowers Association, the Nevada Trappers Association, the Sierra Club, Citizen Alert, the Nevada Miners and Prospectors Association, and many local citizens. The following comments were from a local rancher and a minerals exploration company.

The other range cattle operators and I have lived with and operated according to volumes of rules and regulations to protect the resources and values on our ranges and you propose to take them and blow them to hell. We cannot understand this and especially when you can use the existing facilities without expansion or take the expansion where it won't wreck us.[6]

While I personally am in favor of a strong national defense system, it has in no way been demonstrated that an additional 180,000+ acres in Churchill County is a strategic necessity. I think Nevada has already contributed a disproportional amount of land toward national defense. The military must recognize that with an ever-increasing demand for domestically produced natural resources, it must tailor its land and air requirements to accommodate other equally important uses. On these grounds I am totally opposed to the Navy's withdrawal plan.[7]

Others were more concerned with the planning process.

We believe that a full EIS, which examines a variety of alternatives to the proposal, is necessary. An essential component of an adequate EIS is public information and full public participation on the land withdrawal question.[8]

Without everyone's input, oral or written, throughout the entire process any documents which are produced will be viewed by the State as inadequate. Because of the controversial nature of these proposals, the Navy must take extraordinary steps to assure that anyone who desires to be involved is given that opportunity.[9]

At the request of Nevada State Senator Norman Glaser, the Navy agreed to combine its public hearing on the withdrawal process with other related hearings. These included a hearing on the Navy's proposal to establish a supersonic operations area and a state legislature hearing on federal actions in Nevada, including the basing of F/A-18s at Fallon. This turned out to be a major tactical error in the Navy's planning process because from that day forward, the public linked the master land withdrawal with the supersonic operations area.

The hearing was held in Fallon in October 1982. More than 150 people attended and thirty-five testified. The Navy opened its presentation with a statement that the withdrawal would only involve minor restrictions on activity, such as the height of structures. It pointed out that most uses of the land could continue, including livestock grazing, mining, and geothermal exploration. The Navy's consultant stated that draft environmental impact statements on both the supersonic operations area and the withdrawal would be released by March 1983.

While there is an "ironclad law" that opponents turn out at public hearings and supporters usually have other engagements,[10] the Navy hearing provided an indication of the intensity of project opponents. Most of the public testimony was focused on opposition to the supersonic operations area. Of those who testified on the withdrawal, all but one were opposed. They included ranchers, miners, Dixie Valley residents, wildlife supporters, and Native Americans. The lone supporter was a Fallon rancher who believed that no new restrictions on public use would result from the withdrawal. He referred to the Navy keeping its word in relinquishing the Black Rock and Sahwave ranges.

The generally negative reaction to the Navy's proposals raised fears about whether the Navy would remain in Fallon. The business community and other Navy supporters responded with a letter-writing campaign.

I can tell you firsthand that the Navy Base has been responsible for the continued growth of the Fallon area. Over the years
I have seen the city grow from a sleepy little farming community to a prosperous, thriving small city. . . . The loss of the
Navy Base . . . would be devastating to the merchants and in
the long run all citizens of Churchill County.[11]

Our young pilots are the finest and deserve only the finest
facilities and the vitally necessary training. These young people's lives are on the line daily and we'd be selfish if we provided them with anything less.[12]

The preparation of the draft EIS proved far more complex
and time consuming than originally projected. In August 1984, the
Navy requested an extension of the land segregation through an
emergency withdrawal from the BLM. A segregation is a temporary closure to mining. It was scheduled to expire in October 1984,
after which anyone could file nuisance mining claims and greatly
complicate the withdrawal process.

The Bureau of Land Management did not feel that delays in
completing an environmental analysis were sufficient justification
for an emergency withdrawal. It did indicate, however, that it
would support congressional action to extend the segregation.
Congress eventually extended the segregation "until such withdrawal is acted upon by the Congress." Congress was assured,
however, that the Navy would complete its withdrawal process
"as quickly as possible without compromising on thoroughness
and legal compliance requirements."[13]

The Draft Environmental Impact Statement

After many further delays, the Navy released a draft EIS on
the master land withdrawal in February 1985. The EIS process not
only provides the public with the ability to become constructively
involved in the review of military withdrawals, it also provides
many opportunities to defeat, delay, or modify any proposal. This
can be critical when a project has many constituent parts, any
one of which can be challenged and thereby disrupt the overall
program. Therefore, it can be to the advantage of project propo-
nents to piecemeal a project and write separate environmental im-

pact statements on each part. This is done despite Council on Environmental Quality (CEQ) regulations for comprehensive analysis.

To determine the scope of environmental impact statements, agencies shall consider . . . connected actions, which means that they are closely related and therefore should be discussed in the same impact statement. Actions are connected if they:

(i) Automatically trigger other actions which may require environmental impact statements.

(ii) Cannot or will not proceed unless other actions are taken previously or simultaneously.

(iii) Are interdependent parts of a larger action and depend on the larger action for their justification.[14]

These regulations are subject to interpretation. A broad interpretation would lead to huge EISs because every possible slightly related action would have to be thoroughly analyzed. The Navy apparently rejected a broad interpretation which could have increased the risk of delays to critical programs by combining actions. In addition, political timing was also likely a factor. Early in an administration, political power is at its peak and must be exercised rapidly. Successful exercise of power must be focused on those projects that can be quickly approved. The various proposals for the Fallon Naval Air Station were part of the military buildup of the early Reagan administration. Navy planners realized that the momentum for their projects would likely fade as Reagan approached lame-duck status.

The Navy's political prowess should not be underestimated. Over time the Navy has accumulated significant power through tradition and political IOUs. This has allowed it to establish and implement its own goals. Its institutional memory includes a keen appreciation for the use of timing in the political arena. It has shown the ability to use momentum when available, lie low when conditions are not appropriate, and to resurrect projects when the timing is right.[15]

The most time-consuming component of the Navy's program was the master land withdrawal. The withdrawal process included public review of the EIS and required the approval of Congress.

Of the Navy's other projects, basing of F/A-18s at Fallon was an internal Navy decision and establishment of the supersonic operations area was subject only to Federal Aviation Administration (FAA) approval. In general, the FAA approves military requests for air space designations as a matter of course, much as the Department of the Interior approved land withdrawal requests prior to the Engle Act. Therefore, breaking the expansion program into its component parts was an important decision for Navy planners.

Prior to issuing the draft EIS on the withdrawal, the Navy released a separate EIS on the supersonic operations area and a separate environmental analysis on basing the F/A–18 in Fallon. It came under intense public criticism for this approach after the draft EIS on the supersonic operations area was released, particularly in regards to the supersonic operations area's relationship to the land withdrawal. Consequently, the Navy accelerated the release of its draft EIS for the land withdrawal so that it could be reviewed simultaneously with the supersonic operations area document. This shows that the Navy was receptive to public concerns during the EIS process. Still, the Navy contended that it was dealing with a confused public incapable of understanding its procedures.

Despite these attempts to foster effective public communication and constructive public involvement, there are still widespread uncertainties and misconceptions over the scope and the implications of the proposed land withdrawal. Some of the concerns which continue to be expressed appear to stem from a continuing confusion over the difference between the proposed land withdrawal and the proposed Supersonic Operations Area. It is hoped that, by describing the proposed land withdrawal and the proposed Supersonic Operations Area in two separate DEISs, it will be possible to clarify the separate actions and their respective environmental implications.[16]

In order to meet the accelerated schedule, the draft EIS on the master land withdrawal was issued without detailed internal review. As a result, it was of very poor quality. It contained many factual errors and little backup for its conclusion that the withdrawal would have no significant adverse impacts. This was based

on the premise that most existing uses would be allowed to continue on the withdrawn lands.

The EIS was vague about the need for the withdrawal in the
first place. It was unclear about what the proposed action would
involve in terms of land use restrictions. It failed to objectively
consider any alternatives, including the no action alternative as
required by the Council on Environmental Quality. It did not include mineral or cultural resource inventories as required by the
withdrawal regulations. It contained no analysis of the environmental impacts of the proposed action; rather, it analyzed the impacts of activities on lands adjacent to Navy operations. For
example, it stated that the proposed designation of BLM wilderness areas would have a negative impact on Navy use of air space
above the wilderness areas rather than stating that the expansion
of Navy activities would have a negative impact on the wilderness
areas. The reaction to the EIS was predictable: individuals, organizations, and local, state and federal agencies unanimously blasted
the Navy for its poor effort.

The most significant objections came from the Department of
the Interior. The department's comments criticized the EIS for
being out of date and incomplete; it ignored or poorly quantified
many impacts and it demonstrated its "author's lack of appreciation or knowledge of Nevada." The department charged that it
was a "justification for the Navy proposal rather than an analysis
of impacts," and it contained "many inadequacies, contradictions,
obfustications and errors." Parts of the document were vague and
others jumped to conclusions. The department summarized its review as follows:

> The document, from beginning to end, is confusing. There
> are no cross-references, no points thoroughly supported, and
> nothing for the reviewers and decision makers to consider. It
> is not possible to figure out what the Navy wants nor what it
> means to say.
>
> It appears warranted to remind the Navy that particularly
> with an issue as controversial as their Fallon operation, the
> need for a strong, well-documented and well-written EIS is
> imperative to justify the proposed action and satisfy the con-

cern of taxpayers. This is not just a paperwork exercise to meet the requirements of regulations.[17]

Interior advised the Navy to cease any further work on the EIS, conduct appropriate field inventories, develop a clear description of the proposal, and carry out an active public consultation program. It offered the services of the Bureau of Land Management as an agency with jurisdiction by law and special expertise to assist the Navy in preparing an adequate EIS.

State Concerns

Often, the relationship between state governments and the military is defined by pork barrel politics. That is, states generally look at the military as a source of economic growth and view their own purpose as one of winning a defense contract or being selected as the recipient of a defense project. The state of Nevada, on the other hand, having just played a major role in the defeat of the MX missile project, was experienced and capable in opposing military projects it deemed contrary to its interests. It had become well versed in the use of the EIS process to fight military proposals. That process gave the state an ability to influence the decision-making process in a direct and formal manner.[18]

The state submitted twenty-six pages of comments signed by Governor Richard Bryan. Those comments were directed primarily at the proposed buffer zone for the Bravo 16 range, rather than the adequacy of the document. The state believed that continued use of the existing range, let alone expansion of it, would interfere with the urban development of the Fallon area. Located only eight miles from Fallon, the buffer zone was expected to slow residential development and limit civilian access to wildlife and recreation areas, primarily the Sheckler Reservoir. The state cited a 1984 FAA study that recommended relocating the range to a more remote area. It requested that the Navy develop an alternative to relocate Bravo 16. The state also expressed general concerns about military planning coordination.

The State of Nevada has enjoyed a long partnership with the Department of Defense (DOD) and its various agencies. During recent years, however, this relationship has become strained by conflicts between the Department and the State over the control of land and air space resources in Nevada. This conflict has been aggravated, in our view, by the lack of meaningful Defense Department planning, especially the failure to coordinate among the various DOD agencies in their demands for control of additional public land or air space within the State of Nevada.[19]

The state recommended that the Navy take no further action on the withdrawal until it resolved the Bravo 16 issue and completed a more thorough analysis of the other range expansions. It "strongly recommended" that the Navy coordinate any further work on the EIS with the Bureau of Land Management.

Many other agencies, organizations, and individuals commented on the withdrawal EIS. Churchill County was particularly concerned about impacts to the livestock and mining industries and to the county tax base. The Environmental Protection Agency criticized the poorly defined proposed action, lack of alternatives, and the Navy's piecemeal approach.

The Sierra Club argued that the proposal was part of an effort to establish the continental operations range "in piecemeal fashion by completing the planned improvements at Nellis AFB and Hill AFB as supposedly independent actions." It believed that the Fallon Master Plan was the final stage of the original three-part continental operations range plan.[20] Citizen Alert asked what kept the Navy and Air Force from "getting together and working out a long-range and comprehensive plan for their military activities in Nevada."[21] Many commenters, including the Nevada Department of State Lands, were concerned about maintaining existing uses.

An arrangement which allows the BLM to continue to manage the lands on a daily basis, with the Navy having the final determination of what land uses are permissible is no guarantee that existing uses will continue in the area. The Navy could close the area at any time. We note that commanders, operations and equipment will change through the years, and acceptance of existing uses could just as easily change, irrespective of the present intentions of the Navy.[22]

Most of the individual commenters were Fallon residents objecting to the continued use of Bravo 16. One defined the issue as a taking of private property without compensation because Navy operations on Bravo 16 interfered with the use and enjoyment of his property.[23] Another group of commenters represented the livestock industry. Their views ranged from serious concerns from public land ranchers about access to rangelands during the winter grazing periods[24] to frivolous comments from range consultants expressing "confusion" about Navy conclusions that vegetation would become reestablished on the twenty or so disturbed acres (0.01 percent of the withdrawal) closed to livestock grazing.[25]

While the controversy was still focused on the draft EIS for the land withdrawal, the Navy issued a final EIS on the supersonic operations area. It was established in June 1985.

In response to the taking private property without compensation issue, the secretary of the Navy announced a program to purchase 12,000 acres in Dixie Valley below the supersonic operations area and a one-year delay in supersonic flight over the valley. The parcels were purchased under "friendly condemnation" procedures whereby the Navy offered fair market value for the property but did not force the landowner to sell. Many landowners submitted their properties to federal court to decide fair market value. Eventually, nearly all of the parcels were sold to the Navy.

The community of Dixie Valley no longer exists. Long-time residents held a funeral for their community in May 1987, placing mementos of their homes into a child-sized coffin, which they buried in the Dixie Valley Cemetery.[26] Although the Navy provided some compensation, many residents questioned its overall mission. They felt that the Navy should be fighting for the right of individuals to live where they wish rather than taking over private property.

The Navy made a good-faith effort to respond to the concerns expressed in the withdrawal EIS review process. It ceased any further work on the EIS and formally requested the assistance of the Bureau of Land Management as a cooperating agency. On request from a lead agency, any other federal agency with special expertise and jurisdiction is required, by law, to be a cooperating agency. Cooperating agencies are required to make staff available to the lead agency to help it prepare an adequate EIS. A cooperating agency, however, does not have to support the lead agency's proposal.

When the Navy brought the BLM in as a cooperating agency, it began a gradual shift from a closed rational-comprehensive planning process to a more incremental approach. It was characterized by a negotiation process between the agencies. Upon assumption of cooperating agency status in November 1985, the BLM initiated a detailed mineral inventory, resolved conflicts with the state over cultural resource management, and prepared an analysis of public comments on the withdrawal. Negotiations between the Navy and the bureau on the specifics of the proposed action and alternatives were also initiated.

Despite these positive steps, the Navy continued a public relations campaign against its critics. It accused the media of failing to "clearly define the proposals." The Navy charged that news articles made no distinction between air space and land and that the press led the public to believe that the Navy would be "condemning and taking over their homes and ranches." The Navy also provided advice for its antagonists.

> Those with opposing viewpoints must base their opinions and actions on an extensive evaluation of the actual proposal, and not on information received from possibly unreliable sources. For those who must maintain impartiality, namely the news media, the need for accurate reporting of facts and opinions remains paramount.[27]

The Military Lands Withdrawal Act of 1986

While the Navy was working over its critics in Nevada, back in Washington, Congress was providing clear guidelines for development of the proposed action for the master land withdrawal. Public Law 99–606 (the Military Lands Withdrawal Act of 1986) was essentially a catchall act to deal with several withdrawal extensions for existing ranges that had repeatedly failed to clear Congress. The withdrawals were the Bravo 20 and Nellis ranges in Nevada, the Luke Air Force Range in Arizona, the McGregor Army Range in New Mexico, and the Fort Greely and Fort Wainwright Army ranges in Alaska. Together they amounted to 7 million acres. All of the withdrawals were extended for fifteen years.

In reviewing these proposals, Congress readdressed many of the issues involved in the Engle Act debate. That act had been functional for almost thirty years without amendment or revision and needed updating. Public concerns about the bill focused on many of the same issues raised for the master land withdrawal EIS. The Nevada withdrawals were the subject of special attention.

The Committee has been impressed by the extent to which the State of Nevada is affected by the use of its lands and air space for military purposes and for related purposes directly related to national defense.[28]

In House hearings testimony, many commenters, including Nevada Representative Harry Reid, called for the preparation of a cumulative impact report on all military activities in Nevada. Other commenters, including the Sierra Club, the Wilderness Society, and the Natural Resources Defense Council, wanted to have the Department of the Interior's role in jointly managed areas strengthened. The minerals industry suggested that military land withdrawals were having a negative effect on the nation's ability to compete with mineral imports.

In response to these and other concerns, Congress included a number of measures in the Military Lands Withdrawal Act that have altered subsequent withdrawal proposals. It has been the most significant legislation for military range planning since the Engle Act. Among the most significant measures was Section 3, which called for management of the withdrawn lands by the Department of the Interior. It required the department to prepare resource management plans for livestock grazing, wildlife management, recreation, and fire control for the withdrawn lands.

The withdrawal act made management subject to restrictions necessary to permit military use. It gave the military the authority to close the ranges for military operations, public safety, or national security. While this gives the military final authority over all withdrawn lands, it clearly spells out the intent of Congress to give Interior a major role in their management.

The act also requires preparation of environmental impact

statements on all the withdrawals within twelve years. This included the ranges withdrawn in the act *plus* the Bravo ranges and the master land withdrawal if approved.

There was much debate about how Nevada's special situation should be handled. The House Interior Committee appreciated the unusual extent of the military presence in the state. The committee originally proposed to require a cumulative impact EIS on all current and proposed military land and air space uses in Nevada. It intended for the EIS to permit the people of Nevada to understand the effects of military activities on their state in order to enable them to better plan for and deal with those effects.

Nevada Representative Barbara Vucanovich and others strongly opposed the preparation of an EIS for cumulative impacts. They argued that land and air space withdrawals were separate issues and should not be considered in the same analysis. They worried that a statewide EIS would be highly vulnerable to lawsuits and would be expensive to prepare. They believed that the proposal would "broaden the use of the National Environmental Policy Act by requiring that all withdrawals be addressed in one comprehensive EIS" rather than in separate limited EISs. They also feared that the proposal would "represent a serious change in the way we withdraw public land."[29]

A compromise requiring a "Special Nevada Report" was finally reached. It required the secretaries of the Air Force, Navy, and Interior to prepare a joint report to Congress analyzing the cumulative impacts of all land and air space uses by the military in Nevada. A key provision of the requirement was the development of mitigating measures to reduce the effect of military activities in Nevada. Among other issues, it is intended to address any currently withdrawn lands that the military could relinquish. While the Special Nevada Report is intended to analyze impacts, it is not an environmental impact statement subject to the requirements of the National Environmental Policy Act. Citizens will not have the opportunity to file legal challenges to the report.

The Military Lands Withdrawal Act also dealt with the mining issue. It overruled the 1872 Mining Law for military withdrawals by limiting the property rights that miners could obtain on withdrawn land to mineral ores only. The land surface remains in public ownership. This allows mining activity to continue on withdrawn lands subject to control of the surface for military uses.

Revision of the Master Land Withdrawal

The act provided guidelines for the Navy to finally develop a clearly defined proposed action. The Navy developed a zoning proposal for various levels of land use controls.

Zone A lands would require Navy approval before any public access would be allowed. The RAICUZ study identified these lands as having a moderate to high potential for public safety risks from errant bombs. Zone B lands would be open to public access, with restrictions on development. They would comprise the great majority of the withdrawn lands. Zone C lands would be closed to public access. They would be used for further development of electronic warfare sites.

Extended negotiations with the Bureau of Land Management were needed to develop clear proposals for each allowable use. These resulted in a series of compromises.

All non-Navy projects with new structures, such as power lines, would continue to be managed by the bureau. They would be subject to a general height limitation of fifty feet and would require Navy approval. Mining would continue to be managed by the BLM subject to Navy approval of surface operations, structures, and access. Livestock grazing was proposed to continue under BLM management, subject to Navy approval for any range improvements such a fences.

General recreation would continue under bureau management except that major events such as off-road vehicle races would be subject to Navy approval. Finally, the secretary of the Navy would have the authority to close any or all of the withdrawn lands for military operations, public safety, or national security.

The Navy also clearly outlined and analyzed alternatives to the proposed withdrawal for the first time. The alternatives were developed jointly with the Bureau of Land Management based on public comments on the draft EIS. They included an alternative to totally close the withdrawn lands.

A second alternative would relocate part or all of the ranges.

The intent of this alternative was to respond to state and public concerns about moving the Bravo 16 range.

A third alternative would reduce the size of the withdrawal. Based on public comments and additional information from the mineral inventory, it proposed to eliminate the Shoal Sites, areas in the Bravo 17 withdrawal west of State Route 31, the Wonder Mining District, and a county landfill site near Bravo 16.

A fourth alternative proposed using rights-of-way or cooperative agreements instead of a withdrawal. The final alternative was to deny the withdrawal.

In March 1987, the Navy initiated yet another scoping process. It published a public notice of intent to prepare a supplemental draft EIS for the revised withdrawal proposal. Written comments were requested, but no public hearing was held. About twenty comments were submitted, most being in the form of requests for further clarification of proposed restrictions on specific uses such as transmission lines, radios, and livestock fencing. Again, there were many complaints about Bravo 16 and the Navy's piecemeal approach.

In response to these comments and after nearly a year of extensive negotiations, the Navy prepared a completely revised environmental impact statement. This started a lengthy internal Navy approval process that was delayed by turnover of Naval Facilities Engineering Command staff in San Bruno.

The Special Nevada Report

Meanwhile, progress was being made on the Special Nevada Report as required by the Military Lands Withdrawal Act of 1986. It was to have a profound impact on the master land withdrawal process.

The Special Nevada Report was initiated in 1986. The major participants were the Air Force as lead agency and the Navy, Army, Department of Energy, and Bureau of Land Management as cooperating agencies. The Air Force circulated preliminary internal drafts of Special Nevada Report material among the participating agencies and the state government in January 1989. This material outlined major long-term changes for the Fallon Naval Air Station ranges. It included proposals for a new bombing range, a "land-bridge" between the Bravo 17 and Bravo 19 ranges, and additional military uses for the existing ranges and the master land withdrawal.

A new bombing range, Bravo 18, would partially overlie the proposed Bravo 17 expansion in Fairview Valley and occupy most of Bell Flat. It would be used for live bombing and as a target area for sophisticated "stand-off" weapons. These are the "smart bombs" and guided air-to-ground missiles that were used in the Persian Gulf.

The Bravo 17/Bravo 19 land-bridge would connect the two ranges and be contiguous to the new Bravo 18 range. It would provide a corridor for the firing of stand-off weapons between the ranges. The Navy would require control of the ground surface within the weapons' flight path during these operations. Public safety would require the periodic clearance of people from the area. The land-bridge and Bravo 18 would total an additional 240,000 acres.

Long-term proposals for the electronic warfare range included unrestricted laser operations and other intense military uses. The range would have to be cleared periodically for safety reasons. This was a significant change from the proposed action in the master land withdrawal, which provided for open public access to the area at all times.

The Navy included these proposals despite the fact that they were more of a wish list than concrete plans. This showed a good-faith effort by the Navy to comply with the intent of Congress to provide Nevadans with full information about any potential future plans in the Special Nevada Report. Unfortunately for the Navy, the information from the internal draft material was apparently given to various interest groups. This led to a Navy decision to release their proposals to the press in April 1989.

The Navy also decided to include an analysis of the impacts of the new proposals in the master land withdrawal EIS. Council on Environmental Quality regulations require that an analysis of cumulative impacts with reasonably foreseeable future actions be discussed in EISs. The Navy's efforts to revise the withdrawal EIS were sidetracked by the "Ugly Baby" incident at the Fallon Naval Air Station.

NINE

Operation Ugly Baby and Beyond

On October 30, 1989, Fallon NAS staff initiated Operation Ugly Baby, a routine reconnaissance of public lands surrounding the Bravo 17 range. Its purpose was to determine if there were any bombs or other ordnance on public land. They found more than 1,800 inert items and more than 1,000 items of live ordnance, including 1,000-pound bombs.

After notifying the BLM of the situation, they began disposal operations on November 7. Due to the danger involved in moving the bombs, most were exploded on site. An apparent chance observation of this activity by a reporter resulted in a Reno television station's report on November 8 of a secret Navy attempt to cover up the fact that its pilots had been missing their targets for years. The report caused a major uproar, and Nevada Senators Harry Reid and Richard Bryan criticized the operation's secrecy.

The controversy occurred on the heels of a series of safety problems for the Navy. During the previous three weeks, there were ten serious incidents, including the bombing of a Bureau of Land Management campground near the Chocolate Mountain bombing range in southern California. Debris from the bombing injured one camper. Other incidents involved the crash of a Navy fighter into an apartment complex in Georgia, killing two, and a fire aboard an assault ship in Virginia that injured thirty-one. The Navy ordered a forty-eight-hour suspension of all operations worldwide while it reviewed safety procedures.[1]

Meanwhile, Fallon NAS staff responded to the secrecy charge

by stating that information about the bomb clean-up operation was intentionally withheld from the public in order to prevent curiosity seekers. Public safety could have been compromised during the operation if spectators had been attracted. The Navy said that the operation "was working until the news crew showed up."[2] The Navy also said problems could have been prevented if the master land withdrawal had been approved. That reasoning drew sharp criticism from environmental groups.

> We expected, when we heard about all this, the Navy would try to use it as an excuse to expand, but if they can't manage the land they already have, its senseless to give them more.[3]

Governor Miller agreed, saying that they should "stay within the boundaries that were given to them."[4] Still, he requested a closure of contaminated areas until they could be certified safe. The Bureau of Land Management subsequently closed 30,000 acres adjacent to Bravo 16, Bravo 17, and Bravo 19. About 20,000 of these acres were within the boundaries of the master land withdrawal proposal. The fact that 10,000 acres were not included in the withdrawal indicates a serious flaw in Navy planning procedures. The RAICUZ study should have identified all areas with safety problems. A major weakness is that RAICUZ safety zones are determined by the area of armed overflight, rather than on empirical evidence of where bombs actually fall. For this and other reasons, the Department of Defense has been attempting to modify the RAICUZ process. Progress has been slow because of differences of opinion among the armed services.

The closures were authorized under emergency public safety regulations. They are temporary, pending Navy clearance of the areas. Portions are likely to remain in effect until Congress acts on the withdrawal because the Navy cannot guarantee 100 percent clearance of live ordnance given current bomb detection and disposal technology.

In the short term, the controversy generated by the Ugly Baby incident was a major setback to the Navy's master land withdrawal efforts. On the other hand, it achieved partial closure of the area. In the long term, Ugly Baby may be beneficial to the Navy by increasing public awareness of safety problems around its ranges.

As of July 1991, the revised EIS on the master land withdrawal had not been released for public review. The Navy is still working on revisions, including a discussion of the Special Nevada Report proposals. It is also improving its analysis of off-range ordnance drops in the wake of Operation Ugly Baby. Even after a revised draft EIS is released a lengthy process with many opportunities for further delay awaits the Navy.

Once a revised EIS on the withdrawal proposal is released, the Navy must hold a public hearing to invite comments on the proposal and the EIS. Most likely, a public comment period of ninety days will be allowed. The Navy must then respond to those public comments by modifying its proposal, developing new alternatives, modifying its analysis, making factual corrections, or explaining why the comments do not warrant further response. Then it will prepare a final EIS incorporating those changes.

After an acceptable final EIS is prepared, the Navy will release it for a thirty-day public review period. Based on this public review, the Navy and the bureau will each prepare separate records of decision on the proposal. Those records will outline the recommendations that each agency will make to Congress regarding the proposal. In practice, negotiations generally result in agreements on recommendations.

The public may appeal decisions through agency procedures and take them to federal court. Should court action occur it would most likely be focused on the procedural requirements of the Council on Environmental Quality regulations. Agencies are not required to select an environmentally preferable alternative as their decision and cannot be sued on that basis. But they may be sued on the basis of failing to follow proper procedures for analyzing a proposal.

If the process survives the record of decision stage, the Bureau of Land Management and the Navy will draft legislation to implement the withdrawal. This legislation will then be submitted by the secretary of the interior to Congress together with the Department of the Interior's recommendations.

One of the main ideas behind the Engle Act was that Congress would set the terms and conditions of each specific withdrawal after seeking public advice and assistance. Congress is likely to hold extensive hearings of its own on the proposal and may modify the proposed legislation based on those hearings.

If Congress approves the withdrawal and the president signs the withdrawal bill, the Navy and the bureau will prepare a resource management plan for management of the withdrawn lands. The plan will be within the scope of terms and conditions mandated by Congress. In all likelihood, the Navy has at least another three years of battles to fight on the master land withdrawal.

CONCLUSION

The military services have played a major role in land-use planning in Nevada since before the state was admitted to the Union. Early Army expeditions provided invaluable scientific information on the state's geography. Army installations, although temporary, provided protection for settlers and a boost to the state's economy.

The establishment of the Hawthorne Ammunition Depot in 1926 initiated a long-term relationship between the military and the state. World War II brought the most significant expansion of military land uses to date. The Nellis and Fallon air bases have had a major impact on the state's economy and environment. Navy proposals for the Black Rock and Sahwave ranges played a major role in the passage of the Engle Act in 1958. It reasserted congressional control over military land withdrawals, an authority provided to Congress by the Constitution but previously assumed by the executive branch. The Engle Act had the effect of cooling military demands for lands in Nevada for twenty years.

Renewed interest occurred in the seventies with the Air Force's continental operations range proposal. Opposition to that proposal eventually prevailed. It laid the framework for future coalitions of Nevadans opposed to military uses of their lands. An unlikely coalition of environmentalists, ranchers, and miners successfully fought the MX proposal to create a vast network of missile sites throughout the Great Basin of Nevada and Utah. But opponents were not always successful. They lost battles to cancel the Navy's Bravo 20 bombing range and the Air Force's Groom Range.

State officials often opposed federal military expansion in Nevada, but welcomed Aerojet-General Corporation's land swap scheme for private bombing and rocket-testing ranges. The land swap failed to generate promised economic benefits. Its major impact was an increase in "Keep Out" and "No Trespassing" signs in rural Nevada.

The state of Nevada attempted its own withdrawal for nearly 1 million acres in Mineral County. The Nevada National Guard's proposed armored division training range fizzled due to lack of funding when the Cold War ended.

The Fallon Master Land Withdrawal proposal illustrates current military land use planning processes. The Navy initiated the withdrawal in 1979 in response to concerns about potential interference with its training activities from geothermal operations. The Navy ultimately expanded its proposal to withdraw 181,000 acres for buffer zones, electronic warfare facilities, and training grounds at five of the Fallon ranges. The proposal was confirmed in 1982 through internal Navy Range Air Installation Compatibility Use Zone study procedures. The Navy did not provide formal public participation in the withdrawal planning process until 1983, four years after it developed the withdrawal proposal.

Public reaction was generally opposed to the project. Premature release of a draft EIS on the withdrawal strengthened the opposition. Negative comments on the EIS from federal, state, and local agencies, environmental groups, livestock organizations, mining companies, and the general public convinced the Navy to reformulate its proposal and fully analyze alternatives in a completely revised EIS. It was aided considerably by the Military Lands Withdrawal Act of 1986, providing policy guidance for joint military-Department of the Interior management of the withdrawn lands.

Major delays stemming from staff turnover, premature release of Special Nevada Report material, and public controversy over a bomb clearance operation have prevented release of a draft EIS. After more than ten years, the master land withdrawal is still in a preliminary review phase.

Nevada has not been the sole target for military expansion. Recent Department of Defense proposals have included most states in the West. The largest proposal was the Air Force's Saylor Creek expansion for the Mountain Home Air Force Base in southern Idaho. The 1.3-million-acre project included public lands in northeastern Nevada. The National Guard's Montana Training Center project totaled 980,000 acres. In Utah, the Air Force proposed a 450,000-acre electronic warfare range. In California, the Army proposed to expand the Fort Irwin Tank Training Range by 250,000 acres.[1] Except for the Fort Irwin project, these proposals appear to have fallen victim to public opposition and the end of the Cold War. Nevertheless, a change in political circumstances could resurrect them.

It is clear that the Engle Act achieved its primary mission, to reduce the rate of military withdrawal of land from public use. It paved the way to a more open democratic military land use planning process. Yet problems remain, despite improvements to the process in the National Environmental Policy Act of 1969, the Federal Land Policy and Management Act of 1976, and the Military Lands Withdrawal Act of 1986. The withdrawal process is cumbersome and time consuming and it can lead to rancorous debate and ill will between the military and the general public.

If planning is a "school for social learning," then past successes and failures should be analyzed to develop more effective processes.[2] Most local planning departments are characterized by adherence to open rational-comprehensive planning models. They have learned to provide useful avenues for up-front citizen participation. The Navy's planning process remains primarily in a closed rational-comprehensive tradition as practiced in the fifties by transportation planners, a process characterized by faith in technological solutions to land use issues. The RAICUZ process is a prime example: Navy planners develop standardized models of threats to their operations and attempt to apply them to the real world.

If implementation is viewed as the key to any planning process, the master land withdrawal is not a success. After ten years it is still in a preliminary review stage. This was despite assurances to Congress in 1984 that the withdrawal process would be completed as soon as possible. Much of the delay can be attributed to irreconcilable differences of opinion over land use priorities. Yet, several Navy actions contributed to delays, including a planning process that limited up-front public involvement in project development. The Navy was warned of this problem by a Fallon resident in 1982 who said that "without early public involvement, the Navy may get 'locked into' expansion plans that may not be the best alternative for all concerned."[3]

The environmental impact statement process for the land withdrawal provided eventual public input. That process is the most important tool in the progress that has been made in the military planning system. Public involvement mandated by the National Environmental Policy Act has led to increased citizen participation in military decision making. It provides formal and direct access to the decision process. Its influence on improving participatory democracy at all levels of government should not be underestimated. Yet timing problems remain and the NEPA process can provoke rather than enlist the public. A planning process whereby citizen involvement is encouraged at an earlier stage may be preferable.

For the master land withdrawal, formal public review wasn't provided until four years after the proposal had been developed. By that time, both the Navy and its opponents had taken inflexible positions on the withdrawal.

One major problem the Navy faced was a lack of trust. This was the result of an accumulation of incremental decisions over time. Air Force assertions that the continental operations range and the MX missile basing mode were absolutely vital to national defense proved false. Nevadans have remembered the Air Force's decision to close the Nellis Range to livestock grazing and mining during the fifties. Guarantees that new withdrawals are absolutely necessary and will remain open to public use have not been met with total faith.

Errors in dealing with the public have resulted in failures to achieve military objectives. The Air Force's conclusion that there were no unresolved controversies in establishing the continental operations range was based on the assumption that public objections were the result of misconceptions. It was confident that its extensive public relations campaign had cleared up these misconceptions. The Navy has not been quite as confident; it recognized that its public information processes had not caused opposition to dissipate. However, it reacted defensively, blaming the press for its problems. No land use planning process anywhere has been helped by alienating the media.

The Navy's problems have occurred despite having a necessary and popular mission—national defense. During the MX controversy Congress pointed out that the strategic concerns of national defense have precedence over social, economic, and environmental problems. Given that priority, if its land use proposals are truly necessary for the defense of the nation, the Navy should have had an easy time convincing Congress to approve them. The problems encountered in the withdrawal process could lead to the conclusion that at least parts of their proposals may not be necessary.

Elimination of interservice competition is not the answer to additional demands for land by the military. A fiat that such rivalry is unhealthy, unproductive, and unnecessary and will forthwith be eliminated will not happen. Some rivalry is inherent and healthy in terms of competition to develop the best pilot training program.[4] However, improved coordination is desirable and probably can be achieved. The Special Nevada Report is the most likely vehicle for such improvement.

The passage of the Engle Act, the National Environmental Policy Act, the Federal Land Policy and Management Act, and the Military Lands Withdrawal Act of 1986 have empowered the state and the public with a significant level of influence over military land use decisions. Recent governors, particularly Bryan and Miller, have exercised this power. No longer do they feel as limited as Governor Russell did during the Nevada Test Site controversy when he claimed that there was nothing that a governor could do to halt the tests because they were on federal land. The fact that current withdrawal proposals are on federal land assures that they will be subject to at least a minimum level of citizen control as provided by federal law. Nevada has no law comparable to the National Environmental Policy Act for state or private land. The armed services are required under existing federal regulations to fully integrate their EISs with state-required environmental analyses. This would assure Nevada's citizens improved access to military decision making if the state would pass its own environmental policy act.

No longer can the armed services obtain vast areas of the state by merely making out a slip of paper for the Department of the Interior and saying that it is absolutely necessary for national security. Congress now has the authority to set the terms and conditions of each withdrawal. When used wisely, as in the case of the Military Lands Withdrawal Act of 1986, that authority is exercised on the basis of public advice and assistance.

At times, it has appeared that military planners are getting the message about public involvement in a democratic society. A primary premise behind project Nutmeg to establish a continental nuclear testing range was that public opinion would ultimately determine if a site would be established.

The Navy has made some improvements to its planning processes. In theory, it recognizes that it is part of a larger community. It did respond to public criticism about its draft EIS on the withdrawal and is in the process of preparing an entirely new document as requested by the public. The Navy also reacted responsibly in deciding to include a discussion of future plans in the Special Nevada Report and to analyze the cumulative impacts of those plans in the master land withdrawal EIS. Yet, its reaction to coverage of Operation Ugly Baby ("it was working fine until the news crew showed up") indicates that much room for progress remains.

Air Force Colonel Lacy's comments at the Groom Range hearings demonstrate the military's land use planning problems. The colonel stated that it was a uniquely American experience to have the government inform citizens in advance of a proposed action it intends to take. While that is a major step forward from the pre–Engle Act process, citizens now demand a say in the development of proposed actions by their government. The public no longer accepts the idea of the expert planner who knows best what is needed. Military land use planning is too important to be left to military planners.

American society is based on the idea that the people should be as directly involved as possible in government decision making. This idea, participatory democracy, cuts across all levels of government. The national security arena, however, is the largest exception. Clearly, the nation's defense requires more centralized control than most other governmental functions. Improvements, though, may be possible without compromising security needs.

Other federal agencies, including the Bureau of Land Management and the Forest Service, have moved away from simply listening to the public towards actually allowing citizens to influence land use decisions. Both agencies provide for up-front public involvement in identifying issues, developing planning criteria, and formulating alternatives before a preferred course of action is identified. Flaws exist, but the process is a step toward participatory democracy. There is no reason why the armed services couldn't adopt similar procedures without compromising their basic missions.

Local governments have long used citizen boards such as planning commissions to provide for public involvement. The federal land agencies have also used citizen advisory boards to their advantage. Rather than losing control of the decision process, the agencies have been strengthened by involving the public directly in the process. Citizen advisory boards provide excellent opportunities for educating the public about agency needs and generating public support for their projects. They provide an ideal forum for floating trial balloons. Early notice of agency ideas leads to a more efficient planning process and better-informed decisions. Advisory boards may also be established to assist the armed services without jeopardizing national security.

Will the military move towards a more open planning process? Probably not very quickly and not very much more open than at present. Bureaucratic inertia and resistance to change are powerful forces in the military. Participatory democracy is a threat to those in positions of power and authority; yet, public lands are the common legacy of all citizens. Public participation in decisions about those lands is a right, not a privilege, even when national security is at stake. Military planners should welcome that participation.

NOTES

Preface   1. John Walker and David Cowperthwaite, Nevada State Planning Office, Memorandum to Scott Craigie, Chief of Staff, Governor's Office: Upcoming Military Issues in Nevada, March 3, 1989, III 1–2.

One   1. Russell Elliott, *History of Nevada* (Lincoln: University of Nebraska Press, 1973), 43.

2. Michael Brodhead, "Notes on the Military Presence in Nevada: 1843–1988," *Nevada Historical Society Quarterly* (Winter 1989):263. Vol. 32, no. 4.

3. Elbert B. Edwards, *200 Years in Nevada* (Salt Lake City: Publishers Press, 1978), 268.

4. Brodhead, "Notes on the Military Presence," 263.

5. Edwards, *200 Years in Nevada*, 291.

Two   1. U.S. Army Armament, Munitions and Chemical Command, *Installation Profile: Hawthorne AAP*, Brochure, 1988, 3.

2. John McCloskey, *Seventy Years of Griping: Newspapers, Politics, Government* (University of Nevada, Reno, Oral History Program, 1982), 399.

3. U.S. Bureau of Land Management, *Withdrawals and Restorations Prior to the Federal Land Policy and Management Act of October 21, 1976*, 3.

4. Calvin Coolidge, *Executive Order #4531: Nevada* (1926), 1.

5.   Edwards, *200 Years in Nevada*, 275.

6.   U.S. Marine Corps, "Hawthorne Blooms in the Desert," *The Leatherneck* (May 1932), 11.

7.   William Vincent, "Mt. Grant, Wassuk Range Withdrawal" in Citizen Alert, *Comments on the Nevada Report* (1986), 29.

8.   Michael Phillis, "Tons of Munitions Enter Reno on Way to the Gulf," *Reno Gazette-Journal* (February 1, 1991), 1.

9.   Science Applications Corporation, Inc., *Draft Special Nevada Report*, Langley Air Force Base, Virginia, 1990, 4–20.

10.   Science Applications Corporation, *Draft Special Nevada Report*, 4–1.

Three

1.   U.S. Navy, Naval Air Station, Fallon, "History of NAS Fallon" in *Fallon Air Show '89* (1989), 1.

2.   Brodhead, "Notes on the Military Presence," 271.

3.   Joseph A. Fry, "The History of Defense Spending in Nevada: Preview of the MX?" in Francis Hartigan, ed., *MX in Nevada: A Humanistic Perspective* (Reno: Center for Religion and Life, 1980), 41.

4.   Michael Skinner, *Red Flag* (Novato, Calif.: Presidio Press, 1984), 52.

5.   George Malone, U.S. Senate, "Withdrawals of Public Lands for Public Purposes in the 11 Western Public Land States," *Congressional Record—Senate* (March 5, 1956), 3441–54.

6.   U.S. Congress, House, Committee on Interior and Insular Affairs, *Hearings on H.R. 6272—Military Land Withdrawals*, Serial 1, 85th Cong., 1st Sess., 1957, 301.

7.   Fry, "The History of Defense Spending," 39.

8.   U.S. Bureau of Land Management, *Draft Lahontan Resource Management Plan* (Carson City, 1983), 3–28.

9.   Skinner, *Red Flag*, 13.

10.   Ibid., 73.

11. Science Applications Corporation, *Draft Special Nevada Report*, 2–49, 2–50.

12. U.S. Bureau of Land Management, *Draft Nellis Air Force Range Resource Plan and Environmental Impact Statement* (Las Vegas, 1989), 5–12.

13. U.S. Navy, Commandant, 12th Naval District. *Notice Aviso: Range Closure* (San Francisco, 1944), 1.

14. U.S. Navy, *Summary of Important Points: [Black Rock and Sahwave Ranges]* (1955), 1.

15. Cal Bromund, et al., Letter to Nevada State Office, BLM, September 26, 1955, 1.

16. Robert and Margaret Trego, Letter to the Bureau of Land Management, June 6, 1957.

17. Pershing County Chamber of Commerce, "Letter to Senator Malone" in George Malone, U.S. Senate, "Withdrawals of Public Lands for Public Purposes in the 11 Western Public Land States," *Congressional Record—Senate* (March 5, 1956), 3447.

18. U.S. Navy, *Summary of Important Points*, 6.

19. "Survey Planned of Bombing Range," *Reno Evening Gazette* (March 26, 1956).

20. Pershing County Chamber of Commerce, "Letter to Senator Malone," 3447.

21. Cliff Young, U.S. House of Representatives, *Reports from Congress* (October 24, 1956), 2.

22. "Pershing County Residents Unanimously Denounce Navy Plans for Taking Over Land," *Nevada State Journal* (September 9, 1957).

23. J. H. Prater, Testimony in U.S. Bureau of Land Management, *Official Report of Proceedings in the Matter of: Proposed Public Land Order Withdrawing Public Lands in Nevada for Use by the Department of the Navy in Connection with the Black Rock and Sahwave Aerial Gunnery Ranges, Lovelock Nevada 7 June 1957*, Ace Reporting Company (1957), 18.

24. Trego and Trego, Letter to the Bureau of Land Management.

25. Donald M. McAllister, *Evaluation in Environmental Planning: Assessing Social, Economic, and Political Trade-Offs* (Cambridge, Mass.: MIT Press, 1982), 239.

26. Pershing County Chamber of Commerce, "Letter to Senator Malone," 2.

27. William Koegler, Chief, Division of Withdrawals, U.S. Bureau of Land Management, Memorandum to Associate Director, July 5, 1962, 1.

28. Abe Fortas, Acting Secretary of the Interior, Letter to Frank Knox, Secretary of the Navy, April 13, 1944, 2.

29. Ray Bard, Acting Secretary of the Navy, Letter to Abe Fortas, Acting Secretary of the Interior, August 16, 1944, 2.

30. Edward Rowland, Nevada State Director, Bureau of Land Management, Letter to Western Division, Naval Facilities Engineering Command, March 22, 1974, 1.

31. Warren Branscum, Western Division, Naval Facilities Engineering Command, Letter to U.S. Bureau of Land Management, Nevada State Office, May 5, 1980, 1.

32. Science Applications Corporation, *Special Draft Nevada Report*, 3–48.

33. Timothy Christmann, "Strike University," *Naval Aviation News* (July–August 1985): 18–19, vol. 69, no. 5; Tony Holmes, *Superbase 8, Fallon: Supercarrier in the Desert* (London: Osprey Publishing, 1989), 6, 41.

34. A. Costandina Titus, *Bombs in the Backyard* (Reno: University of Nevada Press, 1986), 55.

35. Ibid., 57.

36. Ibid., 74.

37. Mary Ellen Glass, *Nevada's Turbulent '50s* (Reno: University of Nevada Press, 1981), 46.

38. James Hulse, *Forty Years in the Wilderness: Impressions of Nevada 1940–1980* (Reno: University of Nevada Press, 1986), 61.

Four

1. Clair Engle, U.S. House of Representatives, quoted in *Hearings on H.R. 6272*, 42.

2. *Hearings on H.R. 6272*, 94.

3. Clair Engle, U.S. House of Representatives (Congressional Record, 1957, 5512), quoted in Bob Fulkerson, "Is Nevada Becoming a Military Reservation?" *Nevada Public Affairs Review* (1986)1:32. Vol. 181, 1986, no. 1.

4. Samuel T. Dana and Sally K. Fairfax, *Forest and Range Policy*, 2d. ed. (New York: McGraw-Hill Book Company, 1980), 229.

5. *Hearings on H.R. 6272*, 310.

6. John Saylor, U.S. House of Representatives, quoted in *Hearings on H.R. 6272*, 96.

7. U.S. Council on Environmental Quality, *Regulations for Implementing the Procedural Provisions of the National Environmental Policy Act*. Reprint of 40 CFR Parts 1500–1508 (July 1, 1986), 35–36.

8. *Hearings on H.R. 6272*, 322.

Five

1. U.S. Air Force, *Draft Environmental Statement for Proposed Continental Operations Range* (Washington, D.C., 1974), 5–8.

2. U.S. Air Force, *Draft Environmental Statement*, 2–3.

3. Richard Hackman, "Nevadans Here Face Mock Military Attacks in the Future," *Las Vegas Review-Journal* (December 4, 1973), 13.

4. U.S. Air Force, *Draft Environmental Statement*, 1–14.

5. Ibid., 9–1.

6. Frank Hewlett, "AF Programs Will Give Utah 700 New Jobs," *Salt Lake Tribune* (November 15, 1973).

7. "COR to Provide Economic Boost," *Tonopah Times-Bonanza and Goldfield News* (April 12, 1974), 1.

8. Brendan Riley, "Nevada War Range Dropped," *Reno Evening Gazette* (May 29, 1975), 2.

9. U.S. Air Force, *Summary: MX Deployment Area Selection and Land Withdrawal/Acquisition Environmental Impact Statement* (Washington, D.C. 1980), 2.

10.   U.S. Air Force, *Summary: MX Deployment Area Selection*, 5–14.

11.   K. M. Chrysler, "Revolt in the West that Could Kill Supermissile," *U.S. News and World Report* (March 31, 1988), 47–48.

12.   Jon Christensen, "Citizen Alert's Desert Foxes Play Tough Defense," *In These Times* (October 5–11, 1988), 18.

13.   Hulse, *Forty Years in the Wilderness*, 59.

14.   "The West Reacts to MX: Take it Someplace Else," *U.S. News and World Report* (July 13, 1981), 28.

15.   HDR Sciences, *Summary of Scoping for the MX: Deployment Area Selection/Land Withdrawal Environmental Impact Statement.* Prepared for the United States Air Force Ballistic Missile Office, Norton Air Force Base, California, 1980, F–4.

16.   U.S. Congress, House, Committee on Interior and Insular Affairs, *Basing the MX Missile*, Report, 97th Cong., 1st Sess., 1981, 16.

17.   Ibid., 10.

18.   U.S. Air Force, *Draft Environmental Impact Statement, Groom Mountain Range, Lincoln County, Nevada*, Langley Air Force Base, Virginia, 1984, 3–36.

19.   Chris Chrystal, "AF Admits to Illegality of NTS Land Grab," *Las Vegas Sun* (August 7, 1984).

20.   U.S. Air Force, *Draft Environmental Impact Statement*, 1–5.

21.   Jeff Van Ee, Sierra Club, "Letter to Air Force Captain Donald Zona, 12/28/85" in U.S. Air Force, *Final Environmental Impact Statement, Groom Mountain Range, Lincoln County, Nevada*, Langley Air Force Base, Virginia, 1986, 2–28.

22.   Chrystal, "AF Admits to Illegality."

23.   van Ee, "Letter," 2–30.

24.   Cheri Cinkoske, "Letter to Air Force Captain Don-

ald Zona, 12/30/85,'' in U.S. Air Force, *Final Environmental Impact Statement*, 2–36.

25. Melvin Lacy, Colonel, U.S. Air Force, "Opening Statement, Public Hearing for Renewal of the Groom Range Withdrawal, Caliente, Nevada, 11/19/85," in U.S. Air Force, *Final Environmental Impact Statement*, 2–40.

Six

1. Aerojet General Corporation, *Preliminary Draft Environmental Assessment: Florida-Nevada Land Exchange*, (1986), 2.

2. David Koenig, "Reagan Signs Aerojet Land Swap Deal," *Nevada Appeal* (April 1, 1988), C4.

3. "Nevada-Florida Land Exchange Authorization Act of 1988," *Public Law 100-275*, Section 7.

4. Doug McMillan, "Aerospace Firm Leaves Storey an Abundance of Private Land," *Reno Gazette-Journal* (February 15, 1987), 18A.

5. Doug Martin, Nevada Department of Environmental Protection, Interview with the author, June 15, 1990.

6. U.S. Senate, *Congressional Record—Senate* (March 4, 1988), S 1980; David Koenig, "Senate Panel Passes Aerojet Land Swap Bill," *Las Vegas Review-Journal* (February 18, 1988), 1B; and Ralph Wallestrom, Assistant Regional Director—Wildlife Resources, Region 1, U.S. Fish and Wildlife Service, Memo to Regional Director, Region 1, U.S. Fish and Wildlife Service, December 11, 1986, 3.

7. Koenig, "Reagan Signs Aerojet Land Swap Deal," C4.

8. Wallestrom, Memo to Regional Director, 3; and Jeff van Ee, Sierra Club, Letter to Russ Shay et al., May 13, 1987, 2.

9. Aerojet-General Corporation, *Preliminary Draft Environmental Assessment*, 6–10.

10. Laura Wingard, "Aerojet Plans to Try Again to Get Land Swap Through Congress," *Las Vegas Review-Journal* (November 2, 1986), 1B; and Koenig, "Senate Panel Passes Aerojet Land Swap Bill," 3B.

11. Kris Banvard, "Who'll Clean Up Aerojet Pollution?" *The Sacramento Union* (February 21, 1987), A1, 2.

12. Wallestrom, Memo to Regional Director, 13.

13. Bill Vincent, et al., Ad Hoc Committee on Aerojet Nevada, Letter to Senator Bennett Johnson and Representative Bruce Vento, February 13, 1987, 2.

14. van Ee, "Letter to Russ Shay," 1, 2.

15. Laura Wingard, "State Panel Backs off on Environmental Study," *Las Vegas Review-Journal* (September 17, 1987), 2B.

16. Christopher Beall, "Aerojet Land Swap put Through Grinder," *Las Vegas Review-Journal* (August 14, 1987), 1, 4B.

17. Bill Vincent, Citizen Alert, *Testimony Before the Senate Subcommittee on Interior Appropriations, North Las Vegas* (August 8, 1987), 2.

18. Koenig, "Reagan Signs Aerojet Land Swap Deal," C4.

19. *Public Law 100–275.*

20. Lincoln County Treasurer's Office, Interview with the author, June 8, 1990; and Martha Barlow, Mineral County Treasurer, Interview with the author, June 11, 1990.

21. U.S. Department of the Interior, Board of Land Appeals, *Opinion by Administrative Judge Burski: Appeal from a Decision of the Las Vegas District Office, Nevada, Bureau of Land Management, Rejecting Right-of-Way Application, N–41321*, IBLA 85–422, 95 IBLA 53.

22. Douglas McMillan, "National Guard Wants Hawthorne Site to Instruct Entire Division," *Reno Gazette-Journal* (December 28, 1988), C–2.

23. U.S. General Accounting Office, *Army Training: Need to Improve Assessments of Land Requirements and Priorities*, GAO/NSIAD–90–44BR, December 1989, 9; and Nevada Military Department, *Environmental Assess-*

ment: Reserve Component Training Center (Carson City, 1989), 3.

24. "Guard Training Center Proposed in Hawthorne Area," *Mason Valley News* (October 7, 1988), 8.

25. McMillan, "National Guard Wants Hawthorne Site," C–2.

26. Montana Department of Military Affairs, *Montana Training Center* (Helena, 1989), S2.

27. Harry Reid, U.S. Senate, "Letter to Jack McCloskey," *Mineral County Independent-News* (September 27, 1989).

28. McMillan, "National Guard Wants Hawthorne Site," C–1.

29. Ibid., C–2.

30. Richard Bargen, "Gabbs Valley Warfare Training Range?" *Nevada Outdoor Recreation Association Newsletter* (Autumn, 1988):8.

31. McMillan, "National Guard Wants Hawthorne Site," C–2.

32. Jeanne M. Clark and Daniel McCool, *Staking Out the Terrain* (Albany: State University of New York Press, 1985).

33. Ibid., 25.

34. Ibid., 41.

35. Edward Spang, Nevada State Director, U.S. Bureau of Land Management, Letter to Colonel Jerry Bussell, Nevada Military Department, November 24, 1989, 4.

36. Grace Bukowski, *Citizen Alert Comments: Environmental Assessment: Proposed Reserve Component Training Center* (November 1989), 1.

37. Ibid., 4.

38. Dennis Farney, "Army's Lust for Land, Including Some Where Custer Once Rode, Creates Farmers' Last Stand," *Wall Street Journal* (December 29, 1989), 1.

39. Nevada Military Department, *Environmental Assessment*, 2.

40. U.S. Congress, House, Committee on Interior and Insular Affairs, *Amendments to Federal Land Policy and Management Act of 1976. (H.R. 828)*, Report, 101st Cong., 1st Sess. (July 11, 1989), 12, 13.

41. Ibid., 30.

42. *U.S. Congressional Record—House* (July 17, 1989), H 3789–H 3791.

43. U.S. General Accounting Office, *Army Training*, 9–12.

44. Robert Fulkerson, Executive Director, Citizen Alert, *Testimony before the Subcommittee on National Parks and Public Lands, Committee on Interior and Insular Affairs, U.S. House of Representatives* (January 3, 1990), 8.

45. Michael Phillis, "Guard Commander: Put Training Center on Hold," *Reno Gazette-Journal* (April 13, 1990), B–1.

Seven

1. Harold Seidman and Robert Gilmour, *Politics, Position, and Power: From the Positive to the Regulatory State* (New York: Oxford University Press, 1986), 173.

2. Franklin Roosevelt quoted in Richard Neustadt, "The Power to Persuade," in David Kozak and James Keagle, eds., *Bureaucratic Politics and National Security: Theory and Practice* (Boulder Colo.: Lynne Rienner Publishers, 1988), 153.

3. John Norton Moore and Robert Turner, *The Legal Structure of Defense Organization*. Memorandum prepared for the President's Blue Ribbon Commission on Defense Management (Washington, D.C., 1986), 17.

4. Adam Yarmonlinsky and Gregory Foster, *Paradoxes of Power: The Military Establishment of the Eighties* (Bloomington: Indiana University Press, 1983), 25.

5. David Clary, *Timber and the Forest Service* (Lawrence: University Press of Kansas, 1986), 154.

6. Ibid., 30.

7. Carl Builder, *The Masks of War: American Military Styles in Strategy and Analysis* (Baltimore: Johns Hopkins University Press, 1989), 25–36.

8. Richard Bryan, U.S. Senate, Interview with the author, December 18, 1989.

9. Yarmonlinsky and Foster, *Paradoxes of Power*, 28.

10. Franklin Spinney, *Defense Facts of Life: The Plans/ Reality Mismatch* (Boulder, Colo.: Westview Press, 1985), 83.

11. William Weida and Frank Gertcher, *The Political Economy of National Defense* (Boulder, Colo.: Westview Press, 1987), 70.

12. Don Bradley, Interview with the author, February 16, 1991.

13. U.S. Navy, *Navy Public Works Management*, Report # NAVEDTRA 10893–C (Washington, D.C. 1980), 4–37.

14. Ibid., 4–39.

15. Ibid., 1–7.

16. Bradley, Interview.

Eight   1. Jim Regan, Churchill County Commissioner, Interview with the author, March 8, 1990.

2. U.S. General Accounting Office, *The Navy Can Reduce Its Stated Requirements for F/A–18 Weapons Tactics Trainers*, Report # GAO/NSIAD–84–84 (1984), 1.

3. U.S. Navy, *Statement of Need and Requirements Analysis for Withdrawal of Fallon Ranges*, Fallon Naval Air Station, Nevada (1980), 14–21.

4. U.S. Navy, *Draft Environmental Impact Statement for the Proposed Master Land Withdrawal: Naval Air Station, Fallon, Nevada*, Western Division, Naval Facilities Engineering Command, San Bruno, California (1985), D–3.

5. Ibid., D–8.

6. Don Coops, Letter to Western Division, Naval Facilities Engineering Command, July 15, 1982, 1.

7. Peter Vikre, ASARCO Incorporated, Letter to Western Division, Naval Facilities Engineering Command, January 24, 1983, 1.

8. Citizen Alert, Letter to District Manager, Bureau of Land Management, January 7, 1983, 2.

9. John Sparbel, Nevada State Planning Coordinator, Hearing Presentation, December 22, 1982, 3.

10. Melvin Levin, *Planning in Government* (Chicago: American Planning Association Planners Press, 1987), 219.

11. B. M. Edge, Jones Boys Motors, Letter to Senator Paul Laxalt, September 28, 1984.

12. Daniel Luke, Letter to Capt. Paul Austin, September 30, 1984.

13. U.S. Congress, Senate, *Congressional Record—Senate* (October 2, 1984), S 12699.

14. U.S. Council on Environmental Quality, National Environmental Policy Act, 31.

15. Lowell L. Klessig and Victor L. Strite, *The ELF Odyssey: National Security Versus Environmental Protection* (Boulder, CO: Westview Press, 1980), 58.

16. U.S. Navy, *Draft Environmental Impact Statement*, 1–5.

17. U.S. Department of the Interior, Letter to Western Division, Naval Facilities Engineering Command, May 9, 1985, 20.

18. Lauren Holland, "The Use of NEPA in Defense Policy Politics: Public and State Involvement in the MX Missile Project," *The Social Science Journal* 21 (July 1984) 3:54.

19. Richard Bryan, Governor, state of Nevada, *Governor's Position on the Proposed Master Land Withdrawal* (May 8, 1985), 1.

20. David Hornbeck, Sierra Club, Letter to Western Di-

vision, Naval Facilities Engineering Command, May 9, 1985, 4.

21.   Citizen Alert, *Comments on the Draft Environmental Impact Statement for the Navy Request to Withdraw 181,322.91 Acres of Federally Owned Land from the Public Domain* (March 25, 1985), 2.

22.   Mike Del Grosso, Land Use Planner, Nevada Department of State Lands, Letter to Nevada State Clearinghouse, May 2, 1985, 2.

23.   James Schneider, Letter to Department of the Navy, Western Division, Naval Facilities Engineering Command, February 5, 1985, 2.

24.   Larry Miller, Letter to Western Division, Naval Facilities Engineering Command, April 4, 1985, 2.

25.   Resource Concepts, Inc., Letter to Western Division, Naval Facilities Engineering Command, May 9, 1985, 2.

26.   Douglas McMillan, "Dixie Valley Residents Say Goodbye to Home," *Reno Gazette-Journal* (May 24, 1987).

27.   Steve Burghardt, Public Affairs Officer, Naval Air Station, Fallon, "The Navy in Rural Nevada," *Nevada Public Affairs Review* (1986)1:36.

28.   U.S. Congress, House, Committee on Interior and Insular Affairs, *Withdrawing and Reserving for the Department of the Navy Certain Public Lands Within the Bravo–20 Bombing Range, Churchill County, Nevada, for Use as a Training and Weapons Testing Area, and for Other Purposes*, Report, 99th Cong., 2nd Sess. (1986), 17.

29.   Ibid., 30

Nine

1.   "Congress Must Conduct Inquiry of Navy Flaws," *Reno Gazette-Journal* (November 15, 1989).

2.   Jim Mitchell, "Live Munitions Found Outside Bombing Area," *Reno Gazette-Journal* (November 10, 1989), 1.

3.   Robert Fulkerson, Executive Director, Citizen Alert, in Jim Mitchell, "Navy to Change Pilot Training Drill to

Prevent Stray Bombs Near Fallon,'' *Reno Gazette-Journal* (November 11, 1989), 1.

4.   Jim Mitchell, ''Navy to Change Pilot Training Drill to Prevent Stray Bombs Near Fallon,'' *Reno Gazette-Journal* (November 11, 1989), 1.

Conclusion

1.   Steve Stuebner, ''U.S. Military Plots Vast Land Coups,'' *High Country News* (February 12, 1989), 10.

2.   William Johnson, *The Politics of Urban Planning* (New York: Paragon House, 1989), 10.

3.   Morris LeFever, Letter to Western Division, Naval Facilities Engineering Command, July 12, 1982, 2.

4.   Bryan, Interview.

SELECTED BIBLIOGRAPHY

Bateman, Michael, and Raymond Riley. *The Geography of Defense*. Totowa, N.J.: Barnes and Noble, 1987.

Beneviste, Guy. *Mastering the Politics of Planning: Crafting Credible Plans and Policies that Make a Difference*. San Francisco: Jossey-Bass, 1989.

Brodhead, Michael. "Notes on the Military Presence in Nevada: 1843–1988." *Nevada Historical Society Quarterly* (Winter 1989):261–77.

Builder, Carl H. *Masks of War: American Military Styles in Strategy and Analysis*. Baltimore: Johns Hopkins University Press, 1989.

Burghardt, Steve, Public Affairs Officer, Naval Air Station, Fallon. "The Navy in Rural Nevada." *Nevada Public Affairs Review* (1986)1:33–36.

Callison, Charles. "Good and Bad Medicine: Land Exchanges in the West." *The Amicus Journal* (Fall 1986):19–21.

Christmann, Timothy. "Strike University." *Naval Aviation News* (July–August 1985):18–19.

Clark, Jeanne M., and Daniel McCool. *Staking Out the Terrain*. Albany: State University of New York Press, 1985.

Clary, David. *Timber and the Forest Service*. Lawrence: University Press of Kansas, 1986.

Dana, Samuel T., and Sally K. Fairfax. *Forest and Range Policy*. 2d ed. New York: McGraw-Hill Book Company, 1980.

de Braga, Marcia. *Dig No Graves*. Sparks, Nev.: Western Printing and Publishing, 1964.

Edwards, Elbert B. *200 Years in Nevada*. Salt Lake City: Publishers Press, 1978.

Elliott, Russell. *History of Nevada*. Lincoln: University of Nebraska Press, 1973.

Frazer, Robert. *Forts of the West*. Norman: University of Oklahoma Press, 1965.

Fulkerson, Robert. "Is Nevada Becoming a Military Reservation?" *Nevada Public Affairs Review* (1986)1:27–32.

Glass, Mary Ellen. *Nevada's Turbulent '50s: Decade of Political and Economic Change*. Reno: University of Nevada Press, 1981.

Hartigan, Francis, ed. *MX in Nevada: A Humanistic Perspective*. Reno: Center for Religion and Life, 1980.

HDR Sciences. *Summary of Scoping for the MX: Deployment Area Selection/ Land Withdrawal Environmental Impact Statement*. Prepared for United States Air Force Ballistic Missile Office, Norton Air Force Base, California, 1980.

Holland, Lauren. "The Use of NEPA in Defense Policy Politics: Public and State Involvement in the MX Missile Project." *The Social Science Journal* 21 (July 1984)3:53–71.

Holmes, Tony. *Superbase 8, Fallon: Supercarrier in the Desert*. London: Osprey Publishing, 1989.

Hulse, James. *Forty Years in the Wilderness: Impressions of Nevada 1940–1980*. Reno: University of Nevada Press, 1986.

Johnson, William. *The Politics of Urban Planning*. New York: Paragon House, 1989.

Joss, John. *Strike: U.S. Naval Strike Warfare Center*. Novato, CA: Presidio Press, 1988.

Klessig, Lowell L., and Victor L. Strite. *The ELF Odyssey: National Security Versus Environmental Protection*. Boulder, Colo.: Westview Press, 1980.

Kozak, David, and James Keagle, eds. *Bureaucratic Politics and National Security: Theory and Practice*. Boulder, Colo.: Lynne Rienner Publishers, 1988.

Levin, Melvin. *Planning in Government*. Chicago: American Planning Association Planners Press, 1987.

McAllister, Donald M. *Evaluation in Environmental Planning: Assessing Social, Economic, and Political Trade-Offs*. Cambridge, Mass.: MIT Press, 1982.

McClendon, Bruce W., and Ray Quay. *Mastering Change: Winning Strategies for Effective City Planning*. Chicago: American Planning Association Planners Press, 1988.

McCloskey, John. *Seventy Years of Griping: Newspapers, Politics, Government*. University of Nevada, Reno, Oral History Program, 1982.

Misrach, Richard, and Myriam Weisang Misrach. *Bravo 20: The Bombing of the American West*. Baltimore: Johns Hopkins University Press, 1990.

Montana Department of Military Affairs. *Montana Training Center*. Helena, 1989.

Moore, John Norton, and Robert Turner. *The Legal Structure of Defense Organization*. Memorandum prepared for the President's Blue Ribbon Commission on Defense Management. Washington, D.C., 1986.

Nevada Military Department. *Environmental Assessment: Reserve Component Training Center*. Carson City, 1989.

Patterson, Edna, Louise Ulph, and Victor Goodwin. *Nevada's Northeast Frontier*. Sparks, Nev.: Western Printing and Publishing, 1969.

Ruhlen, George. "Early Nevada Forts." *Nevada Historical Society Quarterly* 7 (1964) 3-4;1-63.

Science Applications Corporation, Inc. *Draft Special Nevada Report*. Langley Air Force Base, Virginia, 1990.

Seidman, Harold, and Robert Gilmour. *Politics, Position, and Power: From the Positive to the Regulatory State*. New York: Oxford University Press, 1986.

Skinner, Michael. *Red Flag*. Novato, Calif.: Presidio Press, 1984.

So, Frank, and Judith Getzels. *The Practice of Local Government Planning*. Washington, D.C.: International City Management Association, 1988.

Spinney, Franklin. *Defense Facts of Life: The Plans/Reality Mismatch*. Boulder, Colo.: Westview Press, 1985.

Titus, A. Costandina. *Bombs in the Backyard*. Reno: University of Nevada Press, 1986.

U.S. Air Force. *Draft Environmental Statement for Proposed Continental Operations Range*. Washington, D.C., 1974.

———. *Draft Environmental Impact Statement on the MX Deployment Area Selection and Land Withdrawal/Acquisition*. Washington, D.C., 1980.

———. *Summary: MX Deployment Area Selection and Land Withdrawal/Acquisition Environmental Impact Statement*. Washington, D.C., 1980.

———. *Draft Environmental Impact Statement, Groom Mountain Range, Lincoln County, Nevada*. Langley Air Force Base, Virginia, 1984.

———. *Final Environmental Impact Statement, Groom Mountain Range, Lincoln County, Nevada*. Langley Air Force Base, Virginia, 1986.

U.S. Army Armament, Munitions and Chemical Command. *Installation Profile: Hawthorne AAP*. Brochure. 1988.

U.S. Bureau of Land Management. *Draft Lahontan Resource Management Plan*. Carson City, 1983.

———. *Draft Nellis Air Force Range Resource Plan and Environmental Impact Statement*. Las Vegas, 1989.

U.S. Bureau of Land Management and U.S. Air Force. *Final Environmental Statement: Proposed Public Land Withdrawal for Nellis Air Force Bombing Range, Nye, Clark, and Lincoln Counties, Nevada*. Las Vegas, 1981.

U.S. Congress. House, Committee on Interior and Insular Affairs. *Hearings on H.R. 6272—Military Land Withdrawals*. Serial 1, 85th Cong., 1st Sess., 1957.

———. *Basing the MX Missile*. Report, 97th Cong., 1st Sess., 1981.

———. *Withdrawing and Reserving for the Department of the Navy Certain Public Lands Within the Bravo-20 Bombing Range, Churchill County, Nevada, for Use as a Training and Weapons Testing Area, and for Other Purposes*. Report, 99th Cong., 2nd Sess., 1986.

———. *Amendments to Federal Land Policy and Management Act of 1976. (H.R. 828)*. Report, 101st Cong., 1st Sess., 1989.

U.S. Congress. Senate, Subcommittee on Public Lands and Reserved Water. *Hearings on S. 2657–S. 2662—Withdraw and Reserve Certain Public Lands for Military Purposes*. 98th Cong., 2nd. Sess., 1984.

U.S. Council on Environmental Quality. *Regulations for Implementing the Procedural Provisions of the National Environmental Policy Act*. Reprint of 40 CFR Parts 1500–1508. July 1, 1986.

U.S. General Accounting Office. *The Navy Can Reduce Its Stated Requirements for F/A-18 Weapons Tactics Trainers*. Report # GAO/NSIAD-84-84. 1984.

———. *Army Training: Need to Improve Assessments of Land Requirements and Priorities*. GAO/NSIAD-90-44BR. December 1989.

U.S. Navy. *Navy Public Works Management*. Report # NAVEDTRA 10893–C. Washington, D.C., 1980.

———. *Naval Facilities Engineering Command History: 1982*. Report OPNAV 5750–1. Alexandria, Virginia, 1982.

———. *Master Plan: Naval Air Station, Fallon, Nevada*. Western Division, Naval Facilities Engineering Command, San Bruno, California, 1983.

———. *Environmental Assessment: Introduction of F/A–18 Naval Strike Fighter at Naval Air Station, Fallon, Nevada*. Naval Air Systems Command (PMA–265), Fallon, Nevada, October 1984.

————. *Draft Environmental Impact Statement for the Proposed Master Land Withdrawal: Naval Air Station, Fallon, Nevada.* Western Division, Naval Facilities Engineering Command, San Bruno, California, 1985.

————. *Basic Military Requirements.* Report# NAVEDTRA 10054F. Washington, D.C., 1986.

U.S. Navy, Bureau of Yards and Docks. *Building the Navy's Bases in World War II.* Vol. I. Washington, D.C., 1947.

U.S. Navy, Naval Air Station, Fallon. "History of NAS Fallon." *Fallon Air Show '89.* 1989.

U.S. President's Blue Ribbon Commission on Defense Management. *A Quest for Excellence: Final Report to the President.* Washington, D.C., 1986.

Vreeland, Patricia, Hamilton Vreeland, and Thomas Lugaski. "Of Mice, Missiles, and Men: The Ecology of Lone Rock, Nevada." *Nevada Historical Society Quarterly* 25 (Spring 1982):46–52.

Weida, William, and Frank Gertcher. *The Political Economy of National Defense.* Boulder, Colo.: Westview Press, 1987.

Yarmonlinsky, Adam, and Gregory Foster. *Paradoxes of Power: The Military Establishment of the Eighties.* Bloomington: Indiana University Press, 1983.

Young, Cliff, U.S. House of Representatives. *Reports from Congress.* October 24, 1956.

INDEX

Bridger, Captain H. C., 21
Bryan, U.S. Senator (also Governor) Richard, 46,
    51, 103, 111
Bureau of Land Management. *See* U.S. Bureau of
    Land Management

C

Carter, President Jimmy, 43
Chocolate Mountain bombing range, 103
Citizen advisory boards, 113
Citizen Alert, 63–64, 66, 87, 94
Citizen involvement. *See* Public participation
Civil War, 2
Clan Alpine Mountains, 24, 28
Cold War, end of, 66, 67, 109
Competition, interservice: and Department of
    Defense, 68; advantages of, 69–70;
    disadvantages of, 69; duplication of training
    ranges, 70; framework for, 68; problems in
    eliminating, 110
Conflicts, interagency, 26–27
Constitution, U.S., 35, 68
Continental Operations Range, 39–42, 94, 107, 110
Coolidge, President Calvin, 6
Cooper, Forest, 21
Cruise Missile Test Routes, ix, 62
Curtiss-Wright land swap, 51

D

Day, Zimmerman, and Basil Corporation, 8, 52
Defense, Department of: 68, 70; lack of
    coordination among services, 11, 93–94;
    modifies RAICUZ process, 104; nation-wide
    joint use study proposed, 60. *See also*
    Competition, interservice
Desatoya Mountains, 28
Desert National Wildlife Range, 10, 13
Desert tortoise, 14, 53–54
Dixie Valley: electronic warfare range, 81;
    geothermal potential, 82; livestock grazing, 28;
    Navy overflights, 84; Navy purchase of, 95

E

Electronic warfare range: Air Force, x, 109;
    Navy, 28, 78, 81, 101